MORE MATHE

ROSS HONSBERGER

THE
DOLCIANI MATHEMATICAL EXPOSITIONS

Published by
THE MATHEMATICAL ASSOCIATION OF AMERICA

———

The Dolciani Mathematical Expositions

NUMBER TEN

MORE MATHEMATICAL MORSELS

ROSS HONSBERGER
University of Waterloo

Published and Distributed by
THE MATHEMATICAL ASSOCIATION OF AMERICA

The DOLCIANI MATHEMATICAL EXPOSITIONS series of the Mathematical Association of America was established through a generous gift to the Association from Mary P. Dolciani, Professor of Mathematics at Hunter College of the City University of New York. In making the gift, Professor Dolciani, herself an exceptionally talented and successful expositor of mathematics, had the purpose of furthering the ideal of excellence in mathematical exposition.

The Association, for its part, was delighted to accept the gracious gesture iniating the revolving fund for this series from one who has served the Association with distinction, both as a member of the Committee on Publications and as a member of the Board of Governors. It was with genuine pleasure that the Board chose to name the series in her honor.

The books in the series are selected for their lucid expository style and stimulating mathematical content. Typically, they contain an ample supply of exercises, many with accompanying solutions. They are intended to be sufficiently elementary for the undergraduate and even the mathematically inclined high-school student to understand and enjoy, but also to be interesting and sometimes challenging to the more advanced mathematician.

DOLCIANI MATHEMATICAL EXPOSITIONS

Vol. 1. *Mathematical Gems,* Ross Honsberger

Vol. 2. *Mathematical Gems II,* Ross Honsberger

Vol. 3. *Mathematical Morsels,* Ross Honsberger

Vol. 4. *Mathematical Plums,* Ross Honsberger (ed.)

Vol. 5. *Great Moments in Mathematics (Before 1650),* Howard Eves

Vol. 6. *Maxima and Minima without Calculus,* Ivan Niven

Vol. 7. *Great Moments in Mathematics (After 1650),* Howard Eves

Vol. 8. *Map Coloring, Polyhedra, and the Four-Color Problem,* David Barnette

Vol. 9. *Mathematical Gems III,* Ross Honsberger

To
Nancy

CONTENTS

PREFACE

Entitling this book *More Mathematical Morsels* has the advantage of immediately informing anyone who is acquainted with my *Mathematical Morsels* just what kind of book to expect. If it hadn't been for the earlier volume, the present collection might well have been called Gleanings From Crux Mathematicorum, referring to wonderful undergraduate problems journal that was founded in Ottawa, Canada, in 1975 by Leo Sauvé and Fred Maskell. Except for a few miscellaneous cases, I encountered all the problems discussed here in *Crux Mathematicorum*. Of particular interest and value are the many slates of problems given by Murray Klamkin in his regular column "Olympiad Corner" during the years 1979–1986. There is much for every level of mathematics scholar to enjoy in *Crux Mathematicorum*—the examples touched on here represent only a small fraction of the good things to be found there.

In this collection we shall be concerned only with elementary problems, and, of those, only ones that aren't too long or complicated. Needless to say, I hope that each of them contains something exciting—a surprising result, an intriguing approach, a stroke of ingenuity. They are not presented in any attempt to be instructive, but solely in the hope of giving enjoyment—an elegant or ingenious argument can be a beautiful thing.

In vindication of the original contributors to *Crux Mathematicorum*, it should be noted that I have written up these brief essays to suit myself, and have generally raided their work as I pleased; consequently, I have only myself to blame for all the faults in the result.

I remember reading somewhere years ago that J. L. Synge declared that *"The mind is at its best when at play."* I have always felt there might be something to this idea and so I hope that you will approach these vignettes as mathematical entertainment.

A SURPRISING PROPERTY
OF THE INTEGER 11

No matter which 55 positive integers may be selected from $(1, 2, \ldots, 100)$, prove that you must choose some two that differ by 9, some two that differ by 10, some two that differ by 12, and some two that differ by 13, but that you need not have any two that differ by 11.

Solution

The following beautiful proof for the case $n = 9$ sprang to mind right away.

Let the integers selected be $x_1, x_2, \ldots, x_{55}$, where

$$A: \quad 1 \leq x_1 < x_2 < \cdots < x_{55} \leq 100.$$

Adding 9 throughout, we get

$$B: \quad 10 \leq x_1 + 9 < x_2 + 9 < \cdots < x_{55} + 9 \leq 109.$$

There are 55 x's in A and 55 $(x + 9)$'s in B, for a total of 110 positive integers, all ≤ 109. By the pigeonhole principle, then, some two of these numbers must be the same. But no two x's are equal, and this implies that no two $(x + 9)$'s can be equal, either. It must be, therefore, that some x_i equals some $x_j + 9$, and we have

$$x_i - x_j = 9, \quad \text{as required.}$$

For $n = 10$, this approach concerns 110 positive integers $\leq$ 110, and the pigeonhole principle has lost its teeth; and for $n = 12$ and 13 it's even worse. While the pigeonhole principle is undoubtedly the key to the entire question, this particular way of using it is evidently not broad enough for our needs, so let's begin again.

We might start by observing that any string S of $2n$ consecutive integers,

$$S = \{a, a+1, \ldots, a+n-1, a+n, a+n+1, \ldots, a+2n-1\},$$

goes together into n pairs, in each of which the difference between the integers is n:

$$(a, a+n), (a+1, a+n+1), \ldots, (a+n-1, a+2n-1).$$

By the pigeonhole principle, not more than n integers can be chosen from the string without taking both numbers from some pair, thus getting two integers that differ by n. Clearly this holds *a fortiori* when more than n integers are taken from a consecutive string containing *fewer* than $2n$ numbers. With these facts at our disposal, the problem yields easily.

For $n = 9$.   If more than 9 integers are taken from any run of 18 or fewer consecutive integers, then some two will differ by 9. Now the set $(1, 2, \ldots, 100)$ partitions into 6 such strings

$$(1–18), (19–36), (37–54), (55–72), (73–90), (91–100).$$

If 55 numbers are selected, the pigeonhole principle implies that at least one of the 6 strings would have to yield up at least 10 integers because 6 9's are only 54, showing that two integers that differ by 9 cannot be avoided. The other cases are handled similarly.

For $n = 10$.   This time we get the 5 strings

$$(1–20), (21–40), (41–60), (61–80), (81–100) .$$

To get 55 numbers, some string must provide at least 11 integers ($5 \cdot 10$ is only 50), giving a pair differing by 10.

For $n = 12$.   For $n = 12$ the strings are

$$(1–24), (25–48), (49–72), (73–96), (97, 98, 99, 100).$$

Even if all four of the integers in the last string are chosen, at least 51 integers must be taken from the first four strings, implying that at least one of them must yield up at least 13 integers ($4 \cdot 12 = 48$), giving the desired conclusion.

For $n = 13$.   For $n = 13$, we have similarly

$$(1–26), (27–52), (53–78), (79–100)$$

(and $4 \cdot 13 = 52$).

The surprising thing is that $n = 11$ just fits the range $(1, 2, \ldots, 100)$ like a glove. We have the 5 strings

$$(1–22), (23–44), (45–66), (67–88), (89–100),$$

from each of which 11 integers *can* be chosen (none is forced to yield $n + 1 = 12$ integers since $5 \cdot 11 = 55$), namely the consecutive runs

$$(1–11), (23–33), (45–55), (67–77), (89–99),$$

with no two differing by 11. In fact, in this case we can even do without the integer 100.

AN UNEXPECTED EQUALITY

In a figure with a lot of symmetry, it always comes as a great surprise to discover equal elements in unsymmetrical positions. How unexpected it is to learn that, for the 3 common tangents to a pair of circles, as shown, the unsymmetrical parts GE and FH along the internal tangent are always the same length.

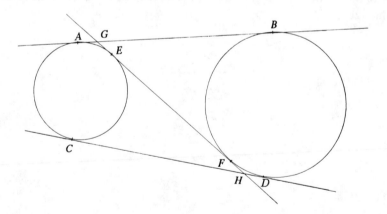

FIGURE 1

Solution

Clearly, by symmetry, the two external tangents AB and CD are equal, and since the two tangents to a circle from the same point are equal, we have

$$AB = AG + GB$$
$$= GE + GF$$
$$= GE + (GE + EF)$$
$$= 2GE + EF;$$

similarly,

$$CD = CH + HD = EH + FH = EF + 2FH.$$

Since $AB = CD$, then $GE = FH$.

From a row of $n \geq 12$ consecutive positive integers, two players, first A and then B, take turns crossing out the integer of their choice until there are just two numbers left, a and b. A wins if a and b are relatively prime, and B otherwise.

   If n is odd, would you choose to play first or second?

   What if n is even?

Solution

(i) Since A wins when a and b are relatively prime, he would be delighted if a and b turned out to be a pair of *consecutive* integers. If n is *odd*, say $2k + 1$, A can always arrange for this to be the case by the following simple strategy.

   On his first move, A strikes the number at one end of the row or the other, leaving an unbroken string of $2k$ consecutive numbers, which clearly go together into k abutting pairs of consecutive integers. A can leave one of these pairs intact at the end by always crossing out the number remaining in the pair that is half-emptied by B on his preceding move.

   (ii) If n is *even*, however, it is better to go second.

   B wins when a and b share any common divisor greater than 1, like 2 or 3. Thus B would be overjoyed to have a and b both even numbers.

If $n = 2k$, there are k odd numbers and k even numbers to begin with, and the game consists of $k - 1$ moves for each player, thus leaving two integers at the end. Hence, if B sticks to striking odd integers, he can get rid of all but one of them on his own. In the event that A helps him in this, by taking even just one odd number, then all the odd numbers would vanish, and the two numbers at the end would both be even, making B the winner. This looks so promising, let us have B adopt this line of play and see how it works. We have seen that A has no option but to counter this by crossing out only *even* integers. Unfortunately for A, B can overcome this defense as follows.

Clearly B wins if a and b are each multiples of 3. Now because $n \geq 12$, the initial row of integers must contain at least 4 multiples of 3; and since the multiples of an odd number alternate between odd and even values, there must be at least 2 *odd* multiples of 3 in the row at the outset, say p and q. Because B is crossing out odd integers, he could inadvertently delete p or q. But if B is careful not to do this, then the final 4 numbers must consist of p, q, and two even integers, unless A has already sealed his doom by taking an odd integer. Since n is even, it would be A's turn when there are just 4 numbers left. And after A's next turn, on which he must continue to take an even number in order to avoid having B leave just the two even numbers, there would be left only one even number with p and q. Thus B can clinch a win by taking the last even number, leaving the multiples p and q.

SANGAKU

During the 18th and 19th centuries it was a custom of Japanese mathematicians to inscribe an especially nice discovery on a tablet and hang
it from the roof of a shrine or temple to the glory of the gods and the
honor of the discoverer. Among such sangaku is the following lovely
result, dated 1893.

Suppose a square piece of paper $ABCD$ is folded by placing the
corner D at some point D' on BC and flattening it out. Suppose AD
is carried into $A'D'$, crossing AB at E. Then, for all choices of D' on
BC, it turns out that the radius of the circle inscribed in $\triangle EBD'$ is just
the length of the overhang $A'E$. (See Figure 2.) Prove this engaging
property.

Solution

The triangle EBD' is right-angled at B. Now, for right triangles there
is a special relation between the lengths of the sides and the size of the
inscribed circle, namely,

$$\text{the diameter} = \text{the sum of the legs} - \text{the hypotenuse.}$$

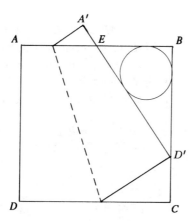

FIGURE 2

This is quite evident from Figure 3, and we shall consider it to be an established result. Accordingly, for $\triangle EBD'$ we have

$$2r = EB + BD' - ED'.$$

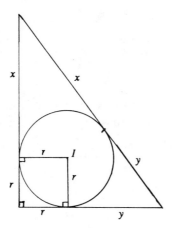

FIGURE 3

Suppose the crease meets AB at F (see Figure 4). Then because $\angle A'$ is a right angle, $\triangle A'FE$ is also right-angled, and since the vertically opposite angles $A'EF$ and BED' are equal, it is similar to $\triangle EBD'$. For some positive constant k, then, each side of $\triangle EBD'$ is k times the corresponding side of $A'EF$, as shown. In these terms our equation becomes

$$2r = kx + ky - kz = k(x + y - z),$$

and

$$r = \frac{k}{2}(x + y - z).$$

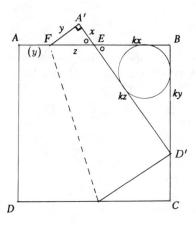

FIGURE 4

Now clearly the folding carries AF into $A'F$, and we have $AF = y$. Then, because the sides AB and $A'D'$ of the square are equal, we have

$$y + z + kx = x + kz,$$

which gives

$$k = \frac{x - y - z}{x - z}.$$

Therefore

$$r = \frac{1}{2} \cdot \frac{x - y - z}{x - z}(x + y - z)$$
$$= \frac{[(x - z) - y][(x - z) + y]}{2(x - z)}$$
$$= \frac{x^2 - 2xz + z^2 - y^2}{2(x - z)}.$$

From the right triangle $A'FE$, however, we have $z^2 - y^2 = x^2$, and hence

$$r = \frac{x^2 - 2xz + x^2}{2(x - z)} = \frac{2x(x - z)}{2(x - z)} = x = A'E,$$

as claimed.

Besides this fine solution I would like to tell you about a most elegant approach that was brought to my attention by Professor Tony Gardiner (University of Birmingham); this is essentially the solution given by Leon Bankoff (Los Angeles) in *Crux Mathematicorum*.

Suppose the side of the given square is s. The key to this solution is the rather unexpected fact that $A'D'$ is always tangent to the quarter-circle having center D and radius s (see Figure 5). To see this, consider the image of the square $ABCD$ when it is reflected in the line of the fold FG. Clearly the images of the three points $A,D,$ and D' of the square are, respectively, the three points A', D', and D. Thus the side AD reflects into $A'D'$, and so the reflection of the opposite side BC will be parallel to $A'D'$; in fact, since D' reflects to D, the image $B'C'$ will lie on the line parallel to $A'D'$ that goes through D. The distance DX across the image-square $A'B'C'D'$ from D to the opposite side $A'D'$ is just the side s, which is the radius of the quadrant, implying that $A'D'$ is indeed a tangent to this arc.

Accordingly, EA and EX are the two tangents from E and we have $AE = EX$; similarly, $D'X = D'C$. Therefore,

$$AE + D'C = EX + XD' = ED'.$$

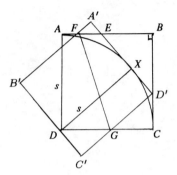

FIGURE 5

But clearly

$$s = AE + EB = D'C + BD' = ED' + A'E;$$

thus $AE = s - EB$, etc., and the relation $AE + D'C = ED'$ yields

$$(s - EB) + (s - BD') = (s - A'E),$$

giving the overhang $A'E$ to be

$$A'E = EB + BD' - s$$

$$= EB + BD' - (ED' + A'E),$$

and so

$$2A'E = EB + BD' - ED'$$

$$= \text{the sum of the legs} - \text{the hypotenuse (of } \triangle EBD')$$

$$= 2r.$$

Hence $r = A'E$, as required.

Exercise. (1982 Alberta High School Prize Examination in Mathematics) A 9×12 rectangular piece of paper is folded so that a pair of diagonally opposite corners coincide. What is the length of the crease?

PAGAN ISLAND

Once upon a time there was an island, called Pagan Island, which had 26 villages around its coastline, whose names in cyclic order were $A, B, C, \ldots, Z$. At various times in its history, the island was visited by 26 missionaries, also named $A, B, C, \ldots, Z$. Each missionary landed on the island at the village which bore his name and he began his work there. To begin with, each village was pagan, of course, but when visited by any missionary, it became converted. Whenever a missionary converted a village, he moved along the coast to the next village in the cycle $A - B - C - \cdots - Z - A$. If a missionary arrived at an unconverted village, he would promptly convert it and continue on his way to the next village in the cycle; however, there was never more than one missionary in a village at any given time. On the other hand, if a missionary arrived at a village that was already in a state of conversion, the villagers, considering themselves put upon too much, would kill the missionary and the village would revert to its pagan condition; this was true even for a missionary who might have gone all the way around the island and revisited a village he originally converted himself. There is no restriction on the number of missionaries that may be on the island at the same time.

After all the missionaries have come and gone, how many of the villages remain converted?

Solution

The island was well named, for after all was said and done, the missionaries brought about no lasting change—all 26 villages were again pagan.

When in a converted state, a village is nothing more than a death-trap for the next unsuspecting missionary. Since a village will be converted by the missionary of the same name, unless it has already been converted by someone else, each village certainly gets converted sometime or other, and thereafter remains a trap until it is sprung on some poor soul. I hope the missionaries weren't told of the conditions in this demonic place, for each of them is doomed to die there; as a missionary goes from village to village, he leaves behind him a string of converted villages, each a trap, the first of which will get him when he comes around again unless he is killed beforehand in a village that was converted by somebody else. Since the villainous village reverts to an unconverted state upon the murder of a missionary, over the years the 26 deaths provide 26 unconverted villages. *If* we could show that once a village has killed a missionary it never gets visited again, and is therefore effectively removed from the proceedings, the next missionary to die must do so in a *different* village, and the 26 deaths ultimately leave behind 26 different unconverted villages. To this end suppose that village A has just murdered its *first* victim; to this point, then, A has been visited by exactly 2 missionaries, the second of whom has just been sent to his reward. Let us consider the possibility of a third missionary visit to A.

One visitor to A might be missionary A when he first arrives on the island, but any other visitor to A must come around the cycle from village Z. If A were to receive 3 visitors, then at least 2 of them would have had to come from Z; that is to say, at least 2 missionaries would actually have made it *into and out of* village Z. Since every second visitor to Z is murdered there, the second one to *leave* Z must have been at least the *third* visitor to Z. Thus, in order for A to receive a third visitor, it is necessary that Z has already had at least 3 visitors. In turn, this requires that Z's predecessor Y must have already had 3 visitors, and so on around the island, until we reach the untenable condition that A, itself, must *previously* have had at least 3 visitors in order to have a third visitor. Our conclusion follows.

PERSISTENT NUMBERS

If a positive integer k contains all ten digits, $0, 1, 2, \ldots, 9$, and this property persists through all its multiples $k, 2k, 3k, \ldots$, then k is said to be a *persistent* number. It turns out that there are no persistent numbers: the demand for persistence through *all* multiples is just too much to ask. An unbroken initial run of successes, however, does not go unrecognized; if the first n multiples $k, 2k, \ldots, nk$, each contain all ten digits, then k is awarded the distinction of being called *n-persistent*. For example, $k = 1234567890$ is 2-persistent because $2k = 2469135780$, but not 3-persistent, in view of $3k = 3703703670$.

Some integers manage to hold out for a long time; the number $k = 526315789473684210$ is 18-persistent, but not 19-persistent.

Prove that there exists at least one n-persistent number for each positive integer n.

Solution

A constructive way of solving this problem would be to come up with a general method of generating an $(n + 1)$-persistent number from an n- persistent number. We shall explain such a procedure and illustrate it by generating a 19-persistent number from the 18-persistent number given above.

The first step is to convert the given n-persistent number k into an n-persistent number that ends in 0. This is accomplished simply by multiplying by 10; a multiple $r(10k)$ of $10k$ contains the same digits as rk except for an extra 0 at the end, and if rk contains all ten digits, so will $r(10k)$. The reason for doing this will become clear in due time; meanwhile, let us proceed with the n-persistent number $N = 10k$.

Of course, it is conceivable that N is also $(n + 1)$-persistent, and so the next thing to do is check out the multiple $M = (n + 1)N$. The problem is how to proceed when M is found lacking.

In this event, our object is to construct an integer P whose multiples $P, 2P, \ldots, (n + 1)P$ each contain all ten digits. We are already in possession of a number N whose multiples $N, 2N, \ldots, nN$ each contain all ten digits. It's just common sense, then, to try to incorporate N into the construction of P so as to have the multiples $N, 2N, \ldots, nN$ convey the ten digits, respectively, into the multiples $P, 2P, \ldots, nP$. This is achieved by building the *right-hand* section of P around N by padding it out on the left with 0's until it extends far enough to accommodate, without overflowing, all $n + 1$ multiples of N.

$$P = \{\ldots \text{left section} \ldots\}\{\ldots \text{right section} \ldots\}$$

$$= \{\ldots \text{left section} \ldots\}\{\underbrace{000 \ldots 0 \overbrace{ab \ldots c}^{N}}_{\text{will hold } (n+1)N}\}.$$

Thus, when P is multiplied by $1, 2, 3, \ldots, n$, this right-hand section will always supply all ten digits on its own, without any help from the left-hand section; the sections do not interfere with each other—that's the purpose of the cushion of 0's. Since N is only n-persistent, this right-hand section *does* need help with the final multiple $(n + 1)P$. This is where the left-hand section comes in. We can arrange for the left section to make up the deficiency in the right section in this case as follows.

Let the digits which are missing from $(n + 1)N$ be listed, in any order, to form an integer Q. The left section of P is to be built around Q so that the result of multiplying this section by $n+1$ will be to produce the number Q. If Q happens to be divisible by $n+1$, so much the better; in general, however, we shall have to adjust it, without disturbing its digits, to make it so. This can be done in the following way. If the number of digits in the divisor $n+1$ is d, then $Q \cdot 10^d$ would simply be the integer Q followed by d 0's:

$$Q \cdot 10^d = \overbrace{st \ldots v}^{Q} \overbrace{00 \ldots 0}^{d \text{ 0's}}.$$

By doing this we have left enough room at the end of this number to add any adjustment $r < n+1$ that might be necessary to make the result divisible by $n+1$, without disturbing the digits of Q at the beginning. Finally, then, P is determined by dividing $Q \cdot 10^d + r$ by $n+1$ and concatenating the integers in the two sections:

$$P = \left\{ \ldots \frac{Q \cdot 10^d + r}{n+1} \ldots \right\} \{00 \ldots 0ab \ldots c\}.$$

For the first n multiples, the left section just goes along for the ride; its sole purpose is to complement the needy right section in the $(n+1)$st multiple.

Now we can see why k was immediately converted into $N = 10k$. Multiples of $10k$ always end in 0, and so never lack the digit 0. In this case Q does not contain 0 and therefore can never reduce to just a single digit 0, which would spoil things by making the whole left section vanish. However, if k itself ends in 0, so will all its multiples, and this preliminary step may be omitted.

This is indeed the case with the given 18-persistent number

$$k = 526315789473684210.$$

Thus we may proceed directly with $N = k$ instead of $N = 10k$. Checking out $19N$, we find

$$19N = 9999999999999999990,$$

and, accordingly, take $Q = 12345678$. Now 19 has 2 digits, making $d = 2$ and, dividing $Q \cdot 10^d$ by $n+1$, we get

$$1234567800 = 19(64977252) + 12;$$

thus the complementary remainder $r = 7$, and we obtain

$$Q \cdot 10^d + r = 1234567807 = 19(64977253),$$

making the left section equal to

$$64977253.$$

In the right section we need to pad the 18-digit N with a single 0 in order to accommodate the 19-digit multiple $(n+1)N = 19N$, which was calculated above. Altogether, then, we obtain the 27-digit 19-persistent number

$$P = 649772530526315789473684210,$$

for which

$$19P = 12345678079999999999999999990.$$

Exercise.   Prove that there are no persistent numbers.

GLEANINGS FROM MURRAY KLAMKIN'S
OLYMPIAD CORNERS—1979

(All references are to the journal *Crux Mathematicorum*.)

1. Olympiad Corner 1, Practice Set (p. 13)

1.   If a positive integer n is composite, then it can be expressed as a product of two factors in at least two different ways:

$$ab = cd \quad (= n)$$

(the factorization $1 \cdot n$ is always available, if desired or needed). For example, 12 is composite and $2 \cdot 6 = 3 \cdot 4$. In this case, the value of

$$S = 2^2 + 6^2 + 3^2 + 4^2 = 65 = 5 \cdot 13.$$

Prove the rather unexpected consequence that, in all cases, the number

$$S = a^2 + b^2 + c^2 + d^2$$

is never a prime number.

At first thought it is very puzzling how the relation $ab = cd$ always makes it possible to factor the integer S. This equality bears on the matter at such a fundamental level that a sophisticated mathematician really has to be quite sharp to see how to use it.

If $ab = cd$, then certainly c divides ab. If this division were to be carried out, suppose that the part of c which divides a is m and the part that divides b is n. In this case, we have

$$c = mn,$$

and for some positive integers p and q, we have

$$a = mp, \qquad b = nq.$$

It follows from $ab = cd$ that

$$d = \frac{ab}{c} = \frac{mp \cdot nq}{mn} = pq.$$

In these terms, we have simply that

$$S = a^2 + b^2 + c^2 + d^2$$
$$= m^2p^2 + n^2q^2 + m^2n^2 + p^2q^2$$
$$= (m^2 + q^2)(n^2 + p^2),$$

a composite number.

It has been observed that this proof goes through unchanged for all powers, not just squares; that is

$$S = a^k + b^k + c^k + d^k$$

is never a prime number.

3.   If a, b, c, are nonnegative real numbers such that

$$(1 + a)(1 + b)(1 + c) = 8,$$

prove that the product abc cannot exceed 1.

It's almost a certainty that the A.M.-G.M. inequality will lie at the center of most solutions, and the problem may therefore be considered fairly routine. However, not every application of this inequality is part of such a pretty approach.

As a first step in marshalling our forces, let's multiply out the given relation to see what we have to deal with:

$$8 = (1+a)(1+b)(1+c)$$

$$= 1 + (a+b+c) + (ab+bc+ca) + abc.$$

From the A.M.-G.M. inequality we get the following line on the term $(a+b+c)$:

$$\frac{a+b+c}{3} \geq \sqrt[3]{abc}.$$

Since the right side is a bit awkward, let's reduce its label to a single symbol, say $\sqrt[3]{abc} = P$. Then

$$a+b+c \geq 3P.$$

The inequality also provides similar information about the term $(ab+bc+ca)$:

$$\frac{ab+bc+ca}{3} \geq \sqrt[3]{ab \cdot bc \cdot ca} = P^2,$$

and

$$ab+bc+ca \geq 3P^2.$$

As a result, the given relation yields the *inequality*

$$8 \geq 1 + 3P + 3P^2 + P^3,$$

which, to our surprise and delight, just happens to be $(1+P)^3$. Accordingly,

$$2 \geq 1 + P, \quad P \leq 1,$$

and we have the desired $P^3 = abc \leq 1$.

Practice Set 3: (page 14)

2. Suppose any selection of n positive integers are lined up in a row. Prove the engaging result that, no matter how wildly they may fluctuate in whatever order you may choose to arrange them, there is always some unbroken block of adjacent integers whose sum is divisible by n. For example: ($n = 9$)

$$71 \quad 7 \quad \underbrace{219 \quad 86 \quad 47 \quad 93 \quad 14}_{\text{sum} = 459 = 9 \cdot 51} \quad 61 \quad 35 \ .$$

Just as surely as the A.M.-G.M. inequality figured to be the key to the last problem, this time there is every likelihood that it will be the famous pigeonhole principle that carries the day.

There are only n possible remainders (pigeonholes) when an integer is divided by n, namely $0, 1, 2, \ldots, n - 1$. While there may be a great many unbroken blocks of adjacent integers in our row, there are precisely n of them which begin with the very first number in the row. If the row consists of the integers $a_1, a_2, \ldots, a_n$, in that order, then these n blocks have sums

$$S_1 = a_1$$

$$S_2 = a_1 + a_2$$

$$S_3 = a_1 + a_2 + a_3$$

$$\vdots$$

$$S_n = a_1 + a_2 + \cdots + a_n.$$

Each of these S_i which is divisible by n is obviously the sum of a desired block. Suppose that none of these S_i is divisible by n. In that case, their n remainders, when divided by n, all lie in the range $\{1, 2, \ldots, n - 1\}$, and the pigeonhole principle asserts that some two of them must be the same. That is to say, for some $i > j$,

$$S_i \equiv S_j \pmod{n}.$$

Accordingly,

$$S_i - S_j \equiv 0 \ (\text{mod } n),$$

that is,

$$a_{j+1} + a_{j+2} + \cdots + a_i \equiv 0 \ (\text{mod } n)$$

and n divides the sum of the block

$$B = \{a_{j+1}, a_{j+2}, \ldots, a_i\}.$$

Let's finish this little section by doing the exercise that was posed in the morsel on persistent numbers.

Exercise.   Prove that there are no persistent numbers.

Recall that a persistent number is a positive integer n all of whose multiples contain all ten digits $0, 1, 2, \ldots, 9$.

If $S = ab \ldots k$ is any string of digits you like, we can prove, by the above approach, that some multiple of any given positive integer n consists of an unbroken block of concatenated strings S followed by a (possibly empty) string of 0's: given any n, there is some m such that

$$mn = SS \ldots S00 \ldots 0$$

$$= ab \ldots kab \ldots k \ldots ab \ldots k00 \ldots 0.$$

Let a_i denote the integer composed of the concatenation of i strings S:

$$a_i = SS \ldots S.$$

Then, if none of the numbers $a_1, a_2, \ldots, a_n$ is itself divisible by n, the pigeonhole principle implies that, for some $i > j$,

$$a_i \equiv a_j (\text{mod } n),$$

$$a_i - a_j = \underbrace{SS \ldots S}_{i-j \text{ times}} \underbrace{0 \ldots 0}_{j \text{ times}} \equiv 0 \ (\mathrm{mod}\ n),$$

establishing our claim.

Consequently, by deliberately omitting some nonzero digit from S, we see that every positive integer n has some multiple $SS \ldots S00 \ldots 0$ which fails to contain all ten digits, implying that sooner or later every positive integer n will prove to be nonpersistent.

2. Olympiad Corner 3, International Olympiad 1977

$S = \{a_1, a_2, \ldots\}$ is a sequence of integers with the unusual property that the sum of any 7 consecutive terms is negative and the sum of any 11 consecutive terms is positive. It seems unlikely that S could carry on like this for very long. What is the maximum length of such a sequence S ?

From the following 11×7 array we can see immediately that S cannot have as many as 17 terms. In the event that S did go to at least 17 terms, each row in the array, consisting of 7 consecutive terms

$$
\begin{array}{cccc}
a_1 & a_2 & \ldots & a_7 \\
a_2 & a_3 & \ldots & a_8 \\
\ldots & \ldots & \ldots & \ldots \\
a_{11} & a_{12} & \ldots & a_{17}
\end{array}
$$

would have a negative sum, implying a negative sum for the entire array. On the other hand, each column consists of 11 consecutive terms, and therefore would have a positive sum, giving the whole array a positive sum, and we have a contradiction. (Such a brilliant observation—unfortunately not mine; I learned of it in a talk by Murray Klamkin.) The remaining task is to see whether we can nail down the desired maximum by actually constructing a sequence S which is as long as possible.

To help us in this undertaking we can deduce the signs of some of the terms by arguing as in the following example: from

$$a_1 + a_2 + a_3 + a_4 = \underbrace{(a_1 + a_2 + \cdots + a_{11})}_{(+\text{ve})} - \underbrace{(a_5 + a_6 + \cdots + a_{11})}_{(-\text{ve})},$$

we get that $a_1 + a_2 + a_3 + a_4$ is *positive*. Similarly,

$$a_4 + a_5 + a_6 + a_7 = (a_4 + \cdots + a_{14}) - (a_8 + \cdots + a_{14})$$

is *positive*. Then

$$a_1 + a_2 + a_3 = \underset{(-ve)}{(a_1 + \cdots + a_7)} \underset{(+ve)}{-(a_4 + \cdots + a_7)}$$

is *negative*. Finally, then,

$$a_4 = \underset{(+ve)}{(a_1 + \cdots + a_4)} \underset{(-ve)}{-(a_1 + a_2 + a_3)}$$

is *positive*. Similar arguments also yield that a_5, a_6, and a_{11} are positive, a_3 and a_7 are negative, and we have enough to attempt a trial construction. With a certain amount of fiddling, one can actually come up with a sequence of length 16, showing the desired maximum is indeed 16; for example,

10, 15, −30, 10, 10, 20, −36, 10, 15, −30, 10, 10, 20, −36, 10, 15.

Exercise.   If every P consecutive terms of S have positive sum and every N consecutive terms have negative sum, prove that S must have fewer than $P + N - D$ terms, where $D = (P, N)$, the gcd of P and N.

3. Olympiad Corner 4, Problem from the journal *Pi Mu Epsilon* 1978 (p. 103)

Some angles—for example, a right angle—can be constructed with the Euclidean tools of straightedge and compasses. Such angles can be repeatedly bisected, duplicated, and combined to form a large set of *constructible* angles; others simply defy construction with Euclidean tools. Similarly, some angles can be *trisected* with straightedge and compasses and others cannot.

The question is, are there angles which, if given, could be trisected with Euclidean tools, but which cannot be constructed with these tools in the first place, that is, are there any angles α which are *trisectible but not constructible?*

The answer is "Yes," for example, $\alpha = 3\pi/7$. If this α were given, then extending either arm would give its supplement $4\pi/7$, and subtraction would yield $\pi/7 = \alpha/3$, thus trisecting α.

It is impossible, however, to construct this angle α with Euclidean tools, for if α could be constructed, we could, by the above, also construct the angle $\beta = 2\alpha/3 = 2\pi/7$, and thereby divide a circle into 7 equal parts (each subtending an angle β at the center), that is, we could construct a regular heptagon. However, this would contradict Gauss' famous theorem that the only regular n-gons it is possible to construct with straightedge and compasses are those for which

$$n = 2^r p_1 p_2 \cdots p_k ,$$

where the p_i are distinct Fermat primes (i.e., $p_i = 2^{(2^m)} + 1$).

4. Olympiad Corner 8, British Olympiad 1979 (p. 227)

3. Suppose S is a set of n *odd* positive integers $a_1 < a_2 < \cdots < a_n$ such that no two of the differences $|a_i - a_j|$ are the same. Prove, then, that the sum Σ of all the integers must be at least $n(n^2 + 2)/3$.

We may proceed by induction. For $n = 1$, the claim holds trivially. Suppose that

$$\Sigma \geq \frac{n-1}{3}[(n-1)^2 + 2]$$

holds for every qualifying set of $n - 1$ integers, and that

$$S = \{a_1 < a_2 < \cdots < a_n\}$$

is a suitable set of n integers.

If no two $|a_i - a_j|$ are equal for i and j in the range $1, 2, \ldots, n$, the same must be true for the subrange $1, 2, \ldots, n - 1$. The induction hypothesis, then, applies to the first $n - 1$ integers, and we have

$$a_1 + a_2 + \cdots + a_{n-1} \geq \frac{n-1}{3}[(n-1)^2 + 2]$$
$$= \frac{n-1}{3}(n^2 - 2n + 3).$$

Now the greatest difference in the set is $a_n - a_1$, implying that

$$a_n = a_1 + (\text{the greatest difference}).$$

Since all the a_i are odd, then each difference is even, placing it among the numbers $\{2, 4, 6, \ldots\}$. But all the differences are *different* values, and the n integers generate $\binom{n}{2}$ of them, and so the greatest difference must be at least as big as the $\binom{n}{2}$th even number, namely $2\binom{n}{2} = n^2 - n$. Thus,

$$a_n \geq a_1 + (n^2 - n) \geq 1 + n^2 - n.$$

Consequently, we have

$$\Sigma = a_1 + a_2 + \cdots + a_n = (a_1 + \cdots + a_{n-1}) + a_n$$
$$\geq \frac{n-1}{3}(n^2 - 2n + 3) + 1 + n^2 - n$$
$$= \frac{1}{3}(n^3 - 2n^2 + 3n - n^2 + 2n - 3 + 3 + 3n^2 - 3n)$$
$$= \frac{1}{3}(n^3 + 2n)$$
$$= \frac{n}{3}(n^2 + 2),$$

and the conclusion follows by induction.

AN APPLICATION OF
van der WAERDEN'S THEOREM

(a) Suppose each point on the circumference of a circle C is colored either red or blue. Prove that, no matter how the colors may be distributed, some 3 points which are *equally spaced* along the circumference will all be the same color.

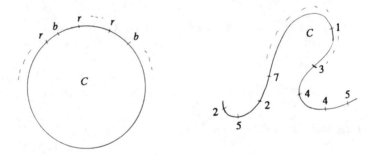

FIGURE 6

(b) More generally, suppose each point on any rectifiable curve C (i.e., distances along the curve are defined) is colored one of k different colors $1, 2, \ldots, k$, where k is any integer > 1. Then, for any integer $r > 2$, prove that some r points that are equally spaced along the curve will all have the same color.

28

Solution

(a)   By the pigeonhole principle, any set of 3 points on the circumference must contain 2 that are the same color. From a set of 3 points that are very near one another, then, we can pick 2 points of the same color that are as close together as we please. Accordingly, let Y and Z be 2 points of the same color that are close enough together so that there is room on the circumference to extend the pair into the ordered quadruple of equally-spaced points (X, Y, Z, W). Let T be the midpoint of the arc YZ, as shown (see Figure 7).

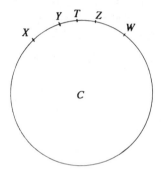

FIGURE 7

   For definiteness, suppose Y and Z are red. Then, if any one of X, T, or W is also red, a trio of equally-spaced red points is determined; otherwise, (X, T, W) itself is a trio of equally-spaced blue points.

(b)   Coloring the points of a curve C with k colors is just a specific way of *partitioning* the points of C into k classes—class i consisting of all the points colored i. Now there is a famous theorem of van der Waerden concerning partitions. It says that, for any given positive integers k and r, there exists a threshold number n, depending on k and r, of course, such that, no matter how the numbers $1, 2, 3, \ldots, n$ may be partitioned into k classes, at least one of the classes will contain an arithmetic progression of length at least r (i.e., having at least r terms). No formula is known for calculating n from k and r; however, the mere existence of such a number n is sufficient for the purpose at hand.

However great n might have to be in order to accommodate the given k and r, one can proceed along the curve C by such small increments so as to determine a set of n equally-spaced points $P_1, P_2, \ldots, P_n$ on C. The k colors partition these n points into k classes, one of which, by van der Waerden's theorem, contains r points whose subscripts are in arithmetic progression:

$$S = \{P_a, P_{a+d}, \ldots, P_{a+(r-1)d}\}$$

(partitioning the set $\{P_i\}$ also partitions the subscripts $1, 2, \ldots, n$). Because the entire set of points $\{P_i\}$ is equally spaced along C, the members of this subset S will also be equally spaced and, belonging to the same class, will all be the same color.

PATRUNO'S PROOF OF
A THEOREM OF ARCHIMEDES

Suppose M is the midpoint of the arc AB of a circle. C is an arbitrary point, other than M, on the arc AMB and D is the foot of the perpendicular from M upon the longer of the chords AC and BC. Prove that D bisects the polygonal path ACB. (This theorem is attributed to Archimedes.)

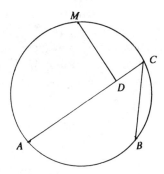

FIGURE 8

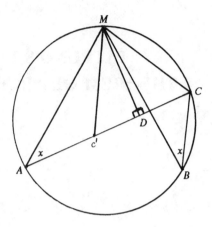

FIGURE 9

Solution

Suppose AC is the longer of the chords AC and BC. Clearly $MA = MB$ since M bisects arc AB; also, chord MC subtends equal angles at A and B at the circumference.

The crux of this marvelous argument, due to Gregg Patruno of New York (a high school student at the time), is the observation that the rotation of $\triangle MBC$ about M, across the circle until MB is carried into MA, causes BC to run along AC, placing C at the point C' such that $AC' = BC$. Thus it remains only to show that $C'D = DC$.

But the rotation also carries MC into MC'. Accordingly, the altitude MD of isosceles $\triangle MC'C$ bisects the base and the proof is complete.

AN INTRACTABLE SUM

What do you do to show that $b > a$ when b is an expression you can't simplify? That's the difficulty in our next problem: for every integer $n > 1$, prove that

$$1 + \frac{1}{2^2} + \frac{1}{3^2} + \cdots + \frac{1}{n^2} > \frac{3n}{2n+1}.$$

Solution

The nature and domain of a variable, though essential in defining the range of validity of a property, often figure only incidentally in the main argument of a solution. We get so used to relegating to the sidelines such qualifying conditions as "for every integer > 1," that we are apt to treat them too lightly when they are of greater importance.

With no formula for the left side of the inequality at hand, we might consider giving more thought to the implication that the domain of n is essentially the set of positive integers. As n runs through $1, 2, 3, \ldots$, the left side generates a *sequence* which we might call

$$\{a_n\}, \qquad \text{where } a_n = \sum_{k=1}^{n} \frac{1}{k^2}.$$

Similarly, the right side generates its own sequence

$$\{b_n\}, \qquad \text{where } b_n = \frac{3n}{2n+1}.$$

The problem can now be interpreted in terms of sequences:

for every integer $n > 1$, prove that $a_n > b_n$.

We can see that this fails for $n = 1$, since both sides are equal to unity; this is undoubtedly why $n = 1$ is not included in the domain of n. However, this little piece of information is not just a useless detail, for it tells us that the two sequences $\{a_n\}$ and $\{b_n\}$ *begin at the same value*. In order to have the required $a_2 > b_2$, then, the sequence $\{a_n\}$ must increase more from a_1 to a_2 than $\{b_n\}$ does from b_1 to b_2. Thus we might be led to an investigation of the relative sizes of the general increments

$$a_n - a_{n-1} \qquad \text{and} \qquad b_n - b_{n-1}.$$

Psychologically we have come a long way from the stunned position of having no formula for a_n.

Accordingly, we have

$$\begin{aligned}
b_n - b_{n-1} &= \frac{3n}{2n+1} - \frac{3(n-1)}{2(n-1)+1} \\
&= 3\left[\frac{n}{2n+1} - \frac{n-1}{2n-1}\right] \\
&= \frac{3}{4n^2-1}.
\end{aligned}$$

For $n > 1$, this fraction is positive but less than 1; and, for such a fraction, the addition of the same positive amount to each of the numerator and denominator causes the value to *increase*. [1] Now then, with a bit of

---

[1] If $0 < x/y < 1$, and $z > 0$, then $y > x$ and

$$\frac{x+z}{y+z} - \frac{x}{y} = \frac{xy + yz - xy - xz}{y(y+z)} = \frac{z(y-x)}{y(y+z)} > 0,$$

since it contains only positive factors.

brilliance, namely the observation

$$b_n - b_{n-1} = \frac{3}{4n^2 - 1} < \frac{3+1}{4n^2} = \frac{1}{n^2} = a_n - a_{n-1},$$

we see that $\{a_n\}$ always increases more from term to term than $\{b_n\}$ does, and that a_n is therefore bound to amount to more than b_n.

A CYCLIC QUADRILATERAL

About a set of four concurrent circles of the same radius r, four of the common tangents are drawn to determine the circumscribing quadrilateral $ABCD$, as shown. Prove that $ABCD$ itself is always a cyclic quadrilateral (i.e., A, B, C, D all lie on a circle).

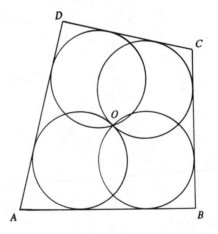

FIGURE 10

Solution

Let the centers of the circles be A', B', C', D', as shown. Since A' and B' are the same distance r from the line AB, $A'B'$ is parallel to AB. Similarly, $A'D'$ is parallel to AD. It follows that the angles at A' and A are equal, and similarly at B' and B, and so on.

Since each center is a distance r from the point of concurrency O, the circle $O(r)$ (i.e., center O, radius r) goes through all of them. Accordingly, the opposite angles A' and C' are supplementary. Thus the identical pair of angles at A and C are also supplementary, and $ABCD$ is cyclic.

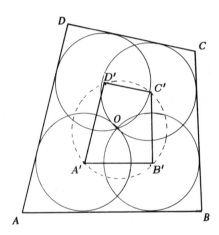

FIGURE 11

Comment. This property does not hold for 5 circles, as can be discovered from the set with centers

$$(1, 1), (-1, 1), (-1, -1), (1, -1), (\sqrt{2}, 0)$$

that are concurrent at the origin.

GLEANINGS FROM MURRAY KLAMKIN'S OLYMPIAD CORNERS—1980

1. Olympiad Corner 11, Practice Set 9 (p. 10)

1.   Suppose each side of quadrilateral $ABCD$ is extended its own length to give 4 new points A', B', C', D', as shown. In the reverse situation, if the points A', B', C', D' were given, how would you construct the quadrilateral $ABCD$?

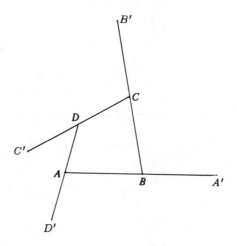

FIGURE 12

This is an ideal place to use vectors. Let an origin O be chosen anywhere in the plane and let $A', B', \ldots$ not only label the given points but also represent the position vectors $OA', OB', \ldots$. Clearly

$$A + \overrightarrow{AB} = B, \qquad \overrightarrow{AB} = B - A,$$

and we have

$$A' = A + 2\overrightarrow{AB} = A + 2(B - A) = 2B - A.$$

In like fashion we can derive the three similar relations

$$B' = 2C - B, \qquad C' = 2D - C, \qquad D' = 2A - D,$$

to complete a set of 4 equations in the 4 unknowns A, B, C, D. The routine solution of these equations yields the vector

$$A = \frac{A' + 2B' + 4C' + 8D'}{15},$$

which is easily constructed.

Once the vertex A is found, the vertex B is located at the midpoint of AA', which leads to C as the midpoint of BB', and finally D as the midpoint of CC'.

2.   The two tangents OA, OB are drawn from a point O to a given circle. Through A a chord AC is drawn parallel to the other tangent OB, and OC crosses the circle at E. Prove the remarkable fact that AE extended always bisects OB.

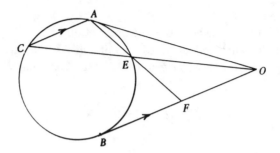

FIGURE 13

The following beautiful solution is the work of Dan Sokolowski of Antioch College.

The two most important basic theorems about tangents, chords, and secants to a circle are the following.

(i)   *The angle between a tangent and a chord through its point of contact is equal to the angle in the segment on the other side of the chord.* Thus we have $\angle OAE = \angle C$.

(ii)   *The product of a whole secant and the part outside the circle is equal to the square on the tangent.* Hence $AF \cdot EF = FB^2$.

Now, because $AC \parallel OB$, the alternate angles ACO and COF are equal. Consequently we have

$$\angle OAF = \angle EOF.$$

Besides this pair of equal angles, in triangles AFO and EFO we also have the common angle at F, implying that the triangles are similar. Therefore

$$\frac{AF}{OF} = \frac{OF}{EF},$$

and we have $OF^2 = AF \cdot EF = FB^2$, giving $OF = FB$, as claimed.

3 (in part).   Prove the standard theorem about a tetrahedron $ABCD$: *if altitudes AX and BY intersect, then the op-*

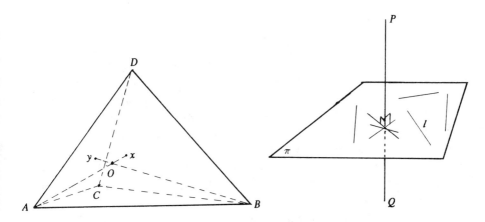

FIGURE 14

posite edges AB and CD, although a pair of skew lines, lie in perpendicular directions.

Suppose AX and BY cross at O. Now, if a straight line PQ is perpendicular to a plane π, then it is perpendicular to every straight line ℓ in π, whether or not they form a pair of skew lines; it is simply a matter of directions, not of concurrency. Thus the vector OA is perpendicular to every straight line in the plane BCD, in particular to CD, and we have that the dot product

$$\overrightarrow{OA} \cdot \overrightarrow{CD} = 0.$$

Similarly $\overrightarrow{OB} \cdot \overrightarrow{CD} = 0$, and by subtraction we have

$$(\overrightarrow{OB} - \overrightarrow{OA}) \cdot \overrightarrow{CD} = 0, \text{ that is } \overrightarrow{AB} \cdot \overrightarrow{CD} = 0,$$

showing AB and CD are perpendicular.

2. Olympiad Corner 13, Practice Set 11 (p. 73)

3.   This problem is being included under false pretenses; I don't intend to discuss the problem at all. It's just that it touches on the wonderful

subject of Dandelin's spheres and I simply can't resist telling you about them.

In the 1820's the Belgian mathematician Germinal Dandelin (1794–1847) discovered a brilliant connection between the synthetic definitions of the conic sections that were given by the ancient Greeks and the more recent analytic definitions involving focal radii. To illustrate his great idea, let's take the case of the ellipse. If an ellipse is defined as the curve of intersection of a right circular cone and a suitably inclined plane, how do you know that there are two points F_1 and F_2 inside the ellipse such that, for every point P on the curve, the sum $PF_1 + PF_2$ is constant? Dandelin's answer is one of the most beautiful little gems of "modern" geometry.

Clearly a sphere S that fits snugly into a right circular cone C makes contact with C in a ring of points R (like a scoop of ice cream.)

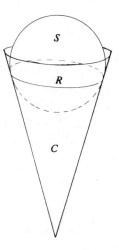

FIGURE 15

Let π be a plane that intersects a right circular cone C, with vertex O, in an ellipse E. Let S_1 and S_2 be the spheres, one on either side of π, that fit snugly into the cone, say along the rings R_1 and R_2, and just make contact with the plane π at single points, say F_1 and F_2 (one can

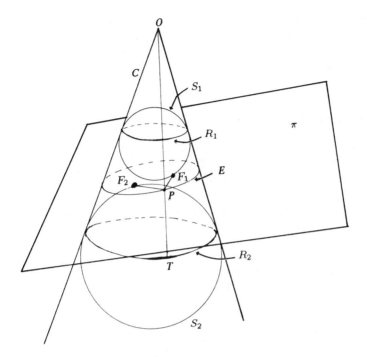

FIGURE 16

imagine small balloons blown up until such snug contacts are achieved).
Let P be any point on E and let the generating line OP cross R_1 at Q
and R_2 at T, as shown.

Then PF_1 and PQ are both tangents to the sphere S_1 from the
point P, PF_1 lying in π and PQ running up the cone. Since the infinity
of tangents to a sphere from a given point are all the same length, we
have

$$PF_1 = PQ.$$

Similarly, $PF_2 = PT$, being equal tangents from P to the sphere S_2.
Thus

$$PF_1 + PF_2 = QP + PT = QT,$$

the distance along a generating line of the cone between the rings of contact R_1 and R_2. But clearly the distance between the rings is the same along all generating lines and we have that $PF_1 + PF_2$ is constant for all points P on E.

3. Olympiad Corner 14, Practice Set 12 (p. 109)

3.   Suppose A and B each flip n fair coins to see who can get more heads; of course, sometimes A wins, sometimes B wins, and occasionally they tie. Naturally, if A is given an extra coin, he has a greater chance of winning. Prove, however, that the advantage of giving an extra coin to A is precisely offset by awarding ties to B.

It seems feasible to argue that A's first n coins only put him on an equal footing with B, and that his chance of winning rests with the extra coin. Since this must come up a head in order to put him over the top, his probability of winning is $1/2$. If ties are given to B, then B wins whenever A doesn't, and B's probability of winning is also $1/2$.

A more rigorous argument is the following. Since A tosses more coins than B, he must get either more heads or more tails, and since he has only *one* extra coin he can't have *both* more heads *and* more tails. Consequently the sum of these probabilities is 1:

Prob(A has more heads) + Prob(A has more tails) = 1.

But these two probabilities are clearly the same for fair coins, and we conclude that each is $1/2$.

Now then, the question I really want to ask is the following.

If A has m coins and B has n, what is the probability that A wins?

The following beautiful solution, based on the symmetry of flipping heads and tails, is due to Basil Rennie of James Cook University of North Queensland, Australia.

In the total collection of outcomes, the number which display a certain preference for heads is surely the same as the number which show the same preference for tails (by the symmetry of fair coins). For example,

$$\text{Prob}(A \text{ has more heads than } B \text{ has heads})$$
$$= \text{Prob}(A \text{ has more heads than } B \text{ has tails})$$

(B is just as likely to get k tails as k heads on any toss)

It follows, then, that

$\text{Prob}(A \text{ gets more heads than } B \text{ does heads})$

$= \text{Prob}(A \text{ gets more heads than } B \text{ does tails})$

$= \text{Prob}(\text{number of } A\text{'s heads} - \text{number of } B\text{'s tails} > 0)$

$= \text{Prob}(\text{number of } A\text{'s heads} - \{n - \text{number of } B\text{'s heads}\} > 0)$

$= \text{Prob}(\text{number of } A\text{'s heads} + \text{number of } B\text{'s heads} > n)$

$= \text{Prob}(\text{ total number of heads in the } m + n \text{ tosses} > n)$

$= \displaystyle\sum_{k=1}^{m} \text{Prob}(\text{total number of heads} = n + k)$

$= \displaystyle\sum_{k=1}^{m} \frac{\binom{m+n}{n+k}}{2^{m+n}}$

$= \displaystyle\frac{1}{2^{m+n}} \sum_{k=1}^{m} \binom{m+n}{n+k}.$

4. Olympiad Corner 17, Canadian Olympiad 1980 (p. 209)

2.   This problem concerns numerical arrays known as Young Tableaux, named after the British clergyman Alfred Young who worked on them around the turn of the century. The subject was touched upon briefly in my guest appearance in Martin Gardner's "Mathematical Games"

column, *Scientific American*, August 1980. As done there, I would like to present the topic in the more striking form of a property of a marching band.

When the leader of a marching band turned to face his musicians he noticed that some of the shorter people were hidden in the pack behind taller players. So he brought the shorter ones forward so that the people in each column appeared in nondecreasing order of height from front to back.

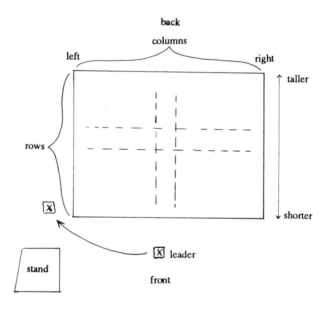

FIGURE 17

Now in the ceremony today, they were to salute the dignitaries in a reviewing stand which they would pass on their right and so the band-master went around to see how they looked from the side. To his dismay he found that some of the shorter players were still blocked from view. To correct this he did to the rows what he had just done to the columns—keeping the rows intact, he arranged the players within each row in nondecreasing order of height from left to right (that is, from *his* left to right as he faced the troop). Naturally he thoroughly expected that this great shuffling inside the rows would completely foul up his carefully ordered columns. You can imagine his surprise and delight when he returned to his post at the front and discovered that the columns were still in nondecreasing order of height from front to back. Prove the remarkable property that such a shuffling of the rows does not spoil the same prearranged order in the columns.

We proceed indirectly. Suppose that after all the rearranging, some player A, say in column i, is closer to the front than a shorter player B. Since the rows were done last, at least we can be sure that the rows are properly ordered. In B's row, then, say row j, we know that nobody in the segment Q up to B is any taller than B himself, and similarly in A's row, say row k, everybody in the segment P from A onwards is at least as tall as A. Since A is taller than B, then everybody in P is taller than everybody in Q.

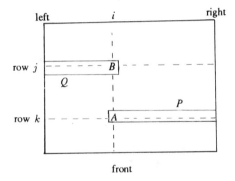

FIGURE 18

Now then, let's back things up to the halfway point when the columns are ordered but the rows still need doing. To achieve this we simply reverse the rearranging that the rows underwent. Suppose the number of columns is n. In row j, then, the members of Q are restored to their former places in the row, that is, they are sorted into the appropriate columns along the row. Similarly, the members of P, while remaining in row k, are also sorted into their former columns. However, from the figure it is obvious that the total number of people in the two segments

Q and P is $n + 1$, exactly one more than the number of columns. According to the pigeonhole principle, then, sorting all these people back into their former columns must cause some two of them to wind up in the same column.

Suppose that some member X of Q occurs in the same column with Y from P. At this halfway point in our 2-step reordering procedure, the columns have already been ordered, and so it must be that X, being farther towards the back (in row j), is at least as tall as Y (in row k). However, since everybody in P is taller than everybody in Q, we also have that Y is taller than X, and the conclusion follows by contradiction.

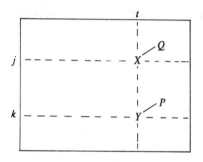

FIGURE 19

5. Olympiad Corner 19, U.S.S.R. National Olympiad 1974 (p. 274)

2. A and B play a game on a given triangle PQR as follows. First A chooses a point X on QR; then B takes his choice of Y on RP, and finally, A chooses Z on PQ. A's object is to make the inscribed $\triangle XYZ$ as large as possible (in area) while B is trying to make it as small as possible. What is the greatest area that A can be sure of getting?

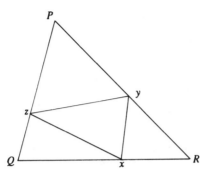

FIGURE 20

If A chooses to make XZ parallel to PR, then all points Y on PR yield a $\triangle XYZ$ with the same area, thus nullifying B's move altogether. If A can't be sure of doing better some other way, at least he can achieve the greatest area subject to $XZ \parallel PR$. Let's determine this maximum.

Suppose X divides QR in the ratio $a:b$. Since all points Y on PR are equivalent, let Y be taken at the foot of the altitude from Q, and suppose QY crosses ZX at S (see figure). Then $\triangle$'s QZX and PQR are similar and

$$\frac{\triangle XYZ}{\triangle PQR} = \frac{\frac{1}{2}ZX \cdot YS}{\frac{1}{2}PR \cdot QY} = \left(\frac{a}{a+b}\right)\left(\frac{b}{a+b}\right) = \frac{ab}{a^2 + 2ab + b^2}.$$

Now, clearly

$$(a - b)^2 \geq 0$$

$$a^2 + 2ab + b^2 \geq 4ab$$

and

$$\frac{ab}{a^2 + 2ab + b^2} \leq \frac{1}{4},$$

with equality if and only if $a = b$.

The maximum is thus $1/4$ and is achieved by taking X and Z to be the midpoints of their sides. Hence, if A can't be sure of doing better, he can always fall back on these midpoints to guarantee $\triangle XYZ = (1/4)\triangle PQR$.

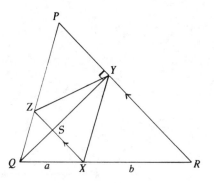

The only catch with this strategy is that A is obliged to commit him-self to it by starting off with X at the midpoint of QR. We can easily see, however, that this is really his only good move anyway, for if A takes X anywhere else on QR, B can turn the tables on him with the same strategy—by choosing Y so as to make XY parallel to PQ. In this way, B nullifies A's second move and, as above, makes

$$\frac{\triangle XYZ}{\triangle PQR} = \frac{ab}{a^2 + 2ab + b^2} < \frac{1}{4}, \qquad \text{since } a \neq b.$$

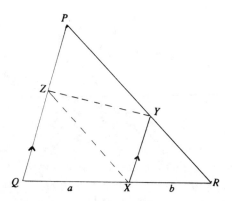

Thus the competitive aspects of this game vanish upon analysis: in every game A would play the midpoints of the sides to obtain $\triangle XYZ = (1/4)\triangle PQR$ since B would punish any deviation from this style of play.

AN UNLIKELY PERFECT SQUARE

When a, b, c are the three rational numbers $\frac{1}{2}, \frac{1}{3}, \frac{1}{5}$, the value of

$$\frac{1}{(a-b)^2} + \frac{1}{(b-c)^2} + \frac{1}{(c-a)^2}$$
$$= \frac{1}{(1/6)^2} + \frac{1}{(2/15)^2} + \frac{1}{(3/10)^2}$$
$$= 36 + \frac{225}{4} + \frac{100}{9}$$
$$= \frac{1296 + 2025 + 400}{36}$$
$$= \frac{3721}{36} = \frac{61^2}{6^2} = \left(\frac{61}{6}\right)^2,$$

is the square of a rational number.

Prove this is no accident, that, in fact, whenever a, b, c are any three different rational numbers, the quantity

$$\frac{1}{(a-b)^2} + \frac{1}{(b-c)^2} + \frac{1}{(c-a)^2}$$

is always the square of a rational number.

Solution

As a small step toward simplifying things, let

$$\frac{1}{a-b} = x, \quad \frac{1}{b-c} = y, \quad \frac{1}{c-a} = z.$$

Then

$$\frac{1}{x} = a - b, \quad \frac{1}{y} = b - c, \quad \frac{1}{z} = c - a,$$

and it is pretty obvious that

$$\frac{1}{x} + \frac{1}{y} + \frac{1}{z} = 0.$$

Clearing of fractions gives

$$yz + zx + xy = 0.$$

Now, the expression we are concerned about is

$$x^2 + y^2 + z^2,$$

which, when mentioned in the same breath with $xy + yz + zx$, can't help but bring to mind the basic expansion

$$(x + y + z)^2 = x^2 + y^2 + z^2 + 2(xy + yz + zx).$$

Since $xy + yz + zx = 0$, this gives

$$x^2 + y^2 + z^2 = (x + y + z)^2,$$

a perfect square!

AN UNLIKELY SYMMETRY

Suppose a circle crosses the sides of a triangle, respectively, in the points $(X, X'), (Y, Y'), (Z, Z')$. If the perpendiculars to the sides at some three of these points happen to be concurrent, prove that the perpendiculars at the remaining three points will also be concurrent.

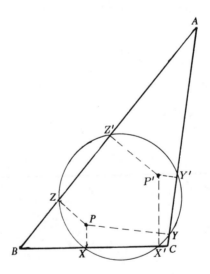

FIGURE 23

54

Solution

At first thought, one might wonder whether there are many circles, if any, that produce this effect. Upon further consideration, however, it is easy to see that there are any number of such circles; in fact, the point of concurrency P can be chosen almost arbitrarily in the plane. From any point P that is not on the circumcircle of the triangle, the perpendiculars PX, PY, PZ to the sides of the triangle determine a suitable circle XYZ (it is not required that the circle *cross* the sides *internally*, or even that it cross a side at all—tangential contacts are permissible; for P on the circumcircle, X, Y, and Z lie on a straight line, and thus fail to determine a circle).

Next, you might expect to be dealing with a rather cluttered situation, what with the triangle, circle, and the six perpendiculars. It could come as a pleasant surprise to discover that the property in question is an immediate consequence of a simple fundamental symmetry which is completely obvious when pointed out.

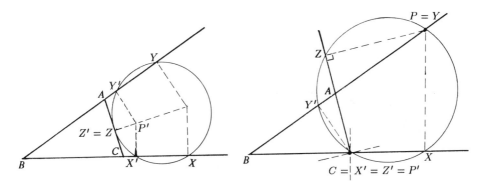

FIGURE 24

Suppose MN is a chord of a circle, center O, and that P is any point on the perpendicular to the chord at M. Let PO meet the perpendicular to the chord at N at the point P'. Now, the perpendicular to the chord of a circle from the center is the perpendicular bisector of the chord—OQ in our figure. As such OQ is the midline of the strip between the parallel perpendiculars at M and N. Thus OQ bisects the transversal PP', and we have the following fundamental symmetry:

> if the perpendicular at one end (M) of a chord of a circle passes through a point P, then the complementary perpendicular at the other end (N) will pass through the point P' which is symmetric to P with respect to the center (O) of the circle.

Accordingly, if a point P lies on three such perpendiculars PX, PY, PZ, then its symmetric point P' will lie on the three complementary perpendiculars.

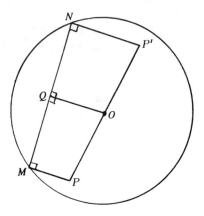

FIGURE 25

TWO FAMOUS DIOPHANTINE EQUATIONS

(a)   Prove that the *only* pair of positive integers (a, b) whose sum equals their product is $(2, 2)$:

$$a + b = ab.$$

(b)   Prove that the *only* pair of *unequal* positive integers (a, b), $a < b$, which satisfy

$$a^b = b^a$$

is $(2, 4)$.

Solution

(a) $a + b = ab$

The solution I am anxious to show you in this section is a gloriously simple and direct one by Sam Greitzer (Rutgers University)—surely "One from the Book"! To set it up, let's first look at several other solutions, each clever in its own right.

(i)   Let b denote the bigger number, if there is one; in all cases, then, we have $a \leq b$.

Now, if $a + b = ab$, then

$$a = \frac{ab}{b} = \frac{a+b}{b} = \frac{a}{b} + 1,$$

which is not an integer for $a < b$. Hence $a = b$, giving

$$2a = a^2, \qquad \text{and} \qquad 2 = a = b.$$

(ii)   Clearly

$$a \mid ab - a.$$

But $ab - a = a + b - a = b$, and we have $a \mid b$.
    Similarly, $b \mid a$, and it follows that $a = b$, etc.

(iii)   As in (i), assign the labels so that $a \leq b$. Since we want $a = 2$, let's investigate the consequences of $a > 2$, i.e., $a \geq 3$. In this case,

$$ab \geq 3b > 2b = b + b \geq a + b = ab,$$

and we have the contradiction $ab > ab$. Thus $a \leq 2$. But $a = 1$ leads to $1 + b = b$, which is impossible. Hence $a = 2$, etc.

(iv)   If $a + b = ab$, then

$$ab - a = b, \quad a(b - 1) = b$$

and, for $b > 1$,

$$a = \frac{b}{b-1} = \frac{(b-1)+1}{b-1} = 1 + \frac{1}{b-1},$$

making $1/(b-1)$ an integer, and this requires $b - 1 = 1$, giving $b = 2$, etc.

And now for Sam's solution.

(v)   If $ab = a + b$, then

$$ab - a - b = 0,$$

$$ab - a - b + 1 = 1,$$

$$(a - 1)(b - 1) = 1,$$

making $a - 1 = b - 1 = 1$, and $a = b = 2$.

(b) $a^b = b^a$

Since $a < b$, let $b/a = 1 + t$, where t is some positive rational number. Then $b = a(1 + t)$ and the equation is

$$a^{a(1+t)} = [a(1 + t)]^a,$$

$$(a^a)^{1+t} = a^a(1 + t)^a,$$

$$(a^a)^t = (1 + t)^a,$$

and, taking ath roots, we get

$$a^t = 1 + t.$$

Now, for positive real numbers t, the exponential series gives

$$e^t = 1 + t + \frac{t^2}{2!} + \cdots > 1 + t.$$

Thus

$$a^t < e^t$$

$$a < e,$$

making $a = 1$ or 2.

If $a = 1$, then, from $a^t = 1 + t$, we must have $t = 0$, a contradiction. Hence the only possibility is $a = 2$.

In this case, the original equation is

$$2^b = b^2, \qquad \text{a perfect square,}$$

showing that the exponent b must be *even*. But $b > a = 2$, and hence the number

$$\frac{b}{2} = \frac{b}{a} = 1 + t$$

must be a positive integer, making t a positive integer. Since $a^t = 1 + t$, we have

$$2^t = 1 + t.$$

It is an easy induction to prove that the *only* solution in positive integers to this equation is $t = 1$: for $t = 1, 2, 3, \ldots$, we have, respectively, $2^t = 2, 4, 8, \ldots$, and $1 + t = 2, 3, 4, \ldots$. Hence $b = 2(1 + 1) = 4$, and the solution $(a, b) = (2, 4)$ is indeed the only one.

BRIANCHON AND CEVA

$ABCD$ circumscribes a given circle S, touching it at the points E, F, G, H, as shown. Let AC and BD cross at I, and let the equal tangents from a vertex have lengths a, b, c, d, respectively, as in the figure. Prove that

$$\frac{AI}{IC} = \frac{a}{c}.$$

Solution

This result follows easy applications of the famous theorems of Brianchon and Ceva.

The setting for Brianchon's theorem (Charles-Julien Brianchon, 1783–1864) is a hexagon which circumscribes a circle; it asserts that the 3 major diagonals (i.e., joining *opposite* vertices) are concurrent. A remarkable characteristic of this theorem is that it holds for its degenerate cases, that is, for pentagons, quadrilaterals, and triangles, in which the vertices of the circumscribing polygon are supplemented by points of contact to make up the 6 points required to identify a hexagon and its diagonals. As a result, it often pays to think of a circumscribing quadrilateral or pentagon as a degenerate hexagon.

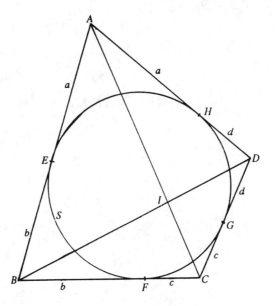

FIGURE 26

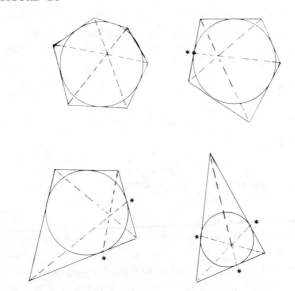

FIGURE 27

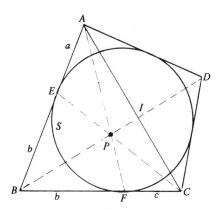

FIGURE 28

Supplementing $ABCD$ with the points of contact E and F, in the present instance we have the circumscribing hexagon $AEBFCD$ and obtain the concurrency of diagonals AF, EC, and BD at some point P.

Now Ceva's theorem (Giovanni Ceva, 1647?–1734) concerns 3 concurrent lines in a triangle, each going from a vertex to the opposite side, like the concurrent AF, BI, and CE in $\triangle ABC$. Ceva's theorem asserts that the product of the ratios into which the sides of the triangle are divided by 3 such lines (called cevians) is always unity. Thus

$$\frac{a}{b} \cdot \frac{b}{c} \cdot \frac{CI}{IA} = 1,$$

and we have

$$\frac{a}{c} = \frac{AI}{IC}.$$

PLAYING ON A POLYNOMIAL

The "board" on which this 2-person game is played is a monic polynomial

$$f(x) = x^{2n} + a_{2n-1}x^{2n-1} + \cdots + a_1 x + 1,$$

in which the $2n - 1$ coefficients $a_{2n-1}, \ldots, a_1$ are unspecified real numbers to begin with. The degree is set at any *even* integer ≥ 4, and the players, starting with A, take turns specifying an undetermined coefficient until $f(x)$ is completely defined. At the end, A wins if no root of $f(x) = 0$ is real, and B otherwise.

Determine a winning strategy for B.

Solution

Since the degree is even, there is always an *odd* number of unspecified terms at the beginning of the game, one more term of odd degree than of even degree. For example, for $2n = 10$, there are 9 unspecified terms, 5 of odd degree (x, x^3, x^5, x^7, x^9) and 4 of even degree (x^2, x^4, x^6, x^8). As we shall see, it is to B's advantage to preserve this delicate imbalance, and so B's initial strategy is simply to restore the status quo by doing just the opposite of what A does—

  if A specifies a term of odd degree,
    then B specifies one of even degree, and vice versa.

At this point it doesn't matter what *value* either A or B may assign to his choice of coefficient. B stays with this approach until there are exactly 3 terms left to be specified. At this stage, then, there must be left two terms of odd degree and one of even degree.

Since there is an odd number of unspecified coefficients to begin with, it is again A's turn to play when there are just 3 terms left, and, after making this move, he must present to B a polynomial with only 2 unspecified terms, ax^s and bx^t, where either

(i) one of s, t is even and the other odd, or

(ii) both s and t are odd.

If $P(x)$ denotes the part of $f(x)$ that has already been specified to this point, then
$f(x) = P(x) + ax^s + bx^t$, where a and b are still to be specified.

Let us consider the two cases separately.

(i)   *Suppose s is even and t is odd.*   In this case,

$$f(1) = P(1) + a + b,$$

$$f(-1) = P(-1) + a - b,$$

giving

$$f(1) + f(-1) = P(1) + P(-1) + 2a.$$

Now B delivers the coup de grâce by choosing the value of a so as to make the right side of this equation equal to zero:

$$a = -\frac{P(1) + P(-1)}{2}.$$

This makes $f(1) + f(-1) = 0$, no matter what A puts for the value of b on the final move of the game. Therefore either

(a) $f(1)$ and $f(-1)$ are themselves both zero, making both $x = 1$ and $x = -1$ real roots, or

(b) one of $f(1), f(-1)$ is negative and the other positive, forcing a real root between 1 and -1 (by the continuity of polynomials).

(ii) *Suppose s and t are both odd.* In this case, $f(1)$ and $f(-1)$ contain the same function of a and b, and cannot be solved for a and b. However, if $f(1)$ is replaced by $f(2)$, we obtain

$$f(-1) = P(-1) - a - b$$
$$f(2) = P(2) + a2^s + b2^t.$$

Thus b is eliminated by multiplying $f(-1)$ by 2^t and adding:

$$2^t f(-1) + f(2) = 2^t \cdot P(-1) + P(2) - a(2^t - 2^s).$$

Again, then, B crushes A by choosing a so as to make the right side vanish:

$$a = \frac{2^t P(-1) + P(2)}{2^t - 2^s}.$$

($s \neq t$ since they are exponents of *different* unspecified terms.) This makes $2^t f(-1) + f(2) = 0$, no matter how A may specify the last coefficient b and again either

(a) both $f(-1)$ and $f(2)$ are themselves zero, or
(b) they straddle zero (the factor 2^t doesn't affect this),

implying a real root in either case.

OVERLAPPING QUADRANTS

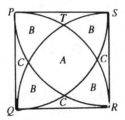

FIGURE 29

Inside the unit square $PQRS$ are drawn the 4 quarter-circles having unit radii and the vertices as centers. Determine the area A that is common to the 4 quadrants.

Solution

There are many ways to solve this old problem—calculus, analytic geometry, trigonometry. The following clever decomposition, however, leads to a most elegant solution.

If the other regions that occur in the figure have areas B and C, as shown, then we immediately obtain the equations

$$A + 4B + 4C = 1, \tag{i}$$

$$A + 3B + 2C = \text{ one quadrant } = \frac{\pi}{4}. \tag{ii}$$

A third equation follows the neat alternative decomposition of the region TQR, of area $A + 2B + C$, that is bounded by the arcs TQ, TR, and the side QR:

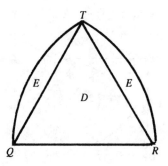

FIGURE 30

D is an equilateral triangle of unit side, and we have

$$D = \frac{1}{2} \cdot 1 \cdot \frac{\sqrt{3}}{2} = \frac{\sqrt{3}}{4};$$

and clearly $D + E$ constitutes a 60° sector of a unit circle, implying

$$D + E = \frac{\pi}{6}.$$

Thus we have

$$A + 2B + C = D + 2E = 2(D + E) - D$$
$$= \frac{\pi}{3} - \frac{\sqrt{3}}{4}. \tag{iii}$$

Solving, we obtain

$$A = 1 - \sqrt{3} + \frac{\pi}{3},$$
$$B = -1 + \frac{\sqrt{3}}{2} + \frac{\pi}{12},$$
$$C = 1 - \frac{\sqrt{3}}{4} - \frac{\pi}{6}.$$

THE CIRCLE AND THE ANNULUS

Let C be a circle of radius 16 and A an annulus having inner radius 2 and outer radius 3. Now suppose that a set S of 650 points is selected inside C. Prove that, no matter how the points of S may be scattered over C, the annulus A can be placed so that it covers at least 10 of the points of S.

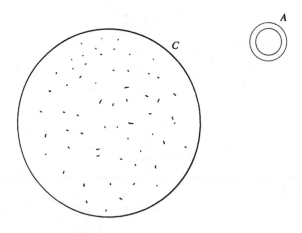

FIGURE 31

Solution

Suppose that a copy of A is centered at each of the 650 points of S. At a point near the edge, the annulus will stick out past the circumference. However, since the center must lie inside C, it can't extend beyond C by as much as its outer radius 3, and therefore a concentric circle D of radius 19 will certainly contain all 650 copies of A in its interior.

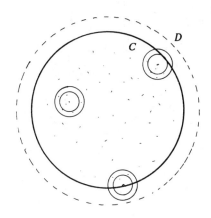

FIGURE 32

Now the area of A is $\pi \cdot 3^2 - \pi \cdot 2^2 = 5\pi$. Hence the 650 copies of A blanket C with a total coverage of

$$650(5\pi) = 3250\pi.$$

Now if no point of D were buried under more than 9 copies of A, then the total area covering D couldn't amount to more than 9 times its area, that is, a total of

$$9(\pi \cdot 19^2) = 9(361\pi) = 3249\pi.$$

A coverage of 3250π, then, must pile up at least 10 copies of A on some point X of D. (This is a *continuous* version of the pigeonhole principle.)

If Y_i is the center of an annulus that covers such a point X, then the distance XY_i must lie between 2 and 3. Because of this, if we turn things around and center a copy $A^\star$ of A at X instead of Y_i, then $A^\star$ would cover Y_i. Since there are at least 10 annuli that cover X, the special annulus $A^\star$, centered at X, then covers their 10 or more centers $Y_1, Y_2, \ldots$, each of which belongs to the given set S.

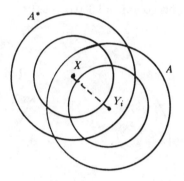

FIGURE 33

GLEANINGS FROM MURRAY KLAMKIN'S
OLYMPIAD CORNERS—1981

1. Moscow Olympiad 1973, Practice Problems (p. 15)

7. Suppose $f(x)$ is a polynomial with integral coefficients. If $f(x) = 2$ for three different integers a, b, and c, prove that $f(x)$ can never be equal to 3 for any integer x.

We begin by establishing the very useful little fact that if $f(x)$ is a polynomial with integral coefficients, then, for any two integers p and q, $f(p) - f(q)$ is *divisible* by $p - q$:

$$p - q | f(p) - f(q).$$

Let $f(x) = c_n x^n + c_{n-1} x^{n-1} + \cdots + c_1 x + c_0$. Then $f(p) - f(q)$ does not contain c_0. Clearly

$$f(p) - f(q) = c_n (p^n - q^n) + c_{n-1}(p^{n-1} - q^{n-1}) + \cdots + c_1(p - q).$$

Since $p - q$ divides each factor $p^m - q^m$ here, the conclusion follows.

Now suppose that, for some integer d, $f(d) = 3$. In view of

$$f(a) = f(b) = f(c) = 2,$$

we would have

$$d - a \mid f(d) - f(a) = 3 - 2 = 1,$$

and similarly, $d - b \mid 1$, $d - c \mid 1$.

Hence each of $d - a$, $d - b$, $d - c$ is either $+1$ or -1, and by the pigeonhole principle some two of them must be the same, making some two of a, b, c the same, a contradiction.

Thus no integer d can make $f(d) = 3$.

2. Moscow Olympiad 1973, Practice Problems (p. 16)

9.   This problem concerns an inkblot on a piece of paper. For each point P of the inkblot the smallest and greatest distances to the edge of the blot are determined: these distances are given by the smallest and greatest radii r of the circles having center P and making contact with the boundary of the inkblot. Then the maximum x of all these smallest distances is determined, and also the minimum y of all these greatest distances.

   If it turns out that $x = y$, what shape must the inkblot have?

Suppose that A is a point whose smallest distance to the edge of the blot is the special value x (presumably there could be more than one such point). Since x is the shortest distance to the boundary from A, the inkblot must completely cover the circle $A(x)$.

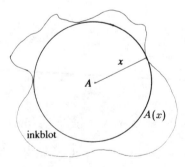

FIGURE 34

Similarly, consider any point B whose greatest distance to the boundary is the special number y. Since circles $B(r)$ with $r > y$ have ceased to make contact with the boundary of the inkblot, it must be that the circle $B(y)$ contains the entire boundary, and thus the entire inkblot. Since the blot contains the circle $A(x)$, then $B(y)$, in covering the blot, must also cover the circle $A(x)$. But $x = y$, and the only way a circle can cover an equal circle is by coinciding exactly with it. In other words, the centers A and B must be the same point, and the circle $A(x)(\equiv B(y))$ both covers and is covered by the inkblot. The inkblot, then, must itself be identical to the circle $A(x)$.

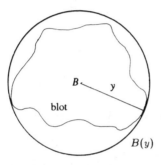

FIGURE 35

3. The Austria-Poland Competition 1980 (p. 42)

8.　S is a set of 1980 points in the plane such that the distance between any two of them is at least 1. Prove that S must contain a subset T of 220 points such that the distance between each two of them is at least $\sqrt{3}$.

We might note that 220 happens to be 1/9th of 1980. You don't suppose there is any chance that each 9 points of S contribute one point to T? We shall see that this is indeed the case; thus our proof will be by induction.

(a)   Let us begin with a set S of 9 points in the plane, each pair of which is at least 1 unit apart. Let O be any point of S on the boundary of the convex hull of S, and let L be a side of the convex hull which goes through O. Then the points of S which don't lie on L all occur on the *same side* of it.

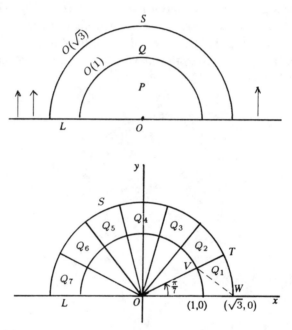

FIGURE 36

On this side of L let the two semicircles $O(1)$ and $O(\sqrt{3})$ determine disjoint regions P and Q, as shown. Now, because no two points of S are closer than 1 unit, none of the other 8 points of S lies inside the region P. We shall show that not more than 7 of them can lie inside the region Q either, implying that at least one point A among the 8 must be at least $\sqrt{3}$ units away from O.

To this end, let Q be partitioned into 7 congruent parts $Q_1, Q_2, \ldots,$ Q_7, by lines radiating from O at angles of $\pi/7, 2\pi/7, 3\pi/7, \ldots, 6\pi/7$ to the line L; also, let cartesian axes be assigned as shown, making O the origin and L the x axis.

Now, it is not quite obvious that the diameter of a region Q_i is the distance VW between a pair of opposite corners (see the figure); conceivably, the chord WT of $O(\sqrt{3})$ could exceed VW. However, if both VW and WT are less than 1, we can be sure that the diameter of Q_i is less than 1. We have, noting that V is $(\cos \pi/7, \sin \pi/7)$ and W $(\sqrt{3}, 0)$,

$$WT < \text{arc } WT = \frac{\text{circumference of } O(\sqrt{3})}{14}$$

$$= \frac{2\sqrt{3}\pi}{14} < .778 < 1; \tag{i}$$

and

$$VW = \sqrt{(\sqrt{3} - \cos \frac{\pi}{7})^2 + \sin^2 \frac{\pi}{7}}$$

$$= \sqrt{3 - 2\sqrt{3} \cdot \cos \frac{\pi}{7} + \cos^2 \frac{\pi}{7} + \sin^2 \frac{\pi}{7}}$$

$$= \sqrt{4 - 2\sqrt{3} \cdot \cos \frac{\pi}{7}} \tag{ii}$$

$$< \sqrt{4 - 2\sqrt{3} \cdot \cos \frac{\pi}{6}} \quad (\text{since } \cos \frac{\pi}{6} < \cos \frac{\pi}{7})$$

$$= \sqrt{4 - 2\sqrt{3} \left(\frac{\sqrt{3}}{2} \right)}$$

$$= 1.$$

Since two points of S are at least 1 unit apart, not more than one could occur in each of the 7 regions Q_i, forcing $OA \geq \sqrt{3}$ for some point A of S.

(b)   Now suppose that, for some $k \geq 1$, a plane set of $9k$ points, each two at least 1 unit apart, contains a set of k points with each two at least $\sqrt{3}$ units apart. Let S be a set of $9k + 9$ points, each two at least 1 unit apart, and let O be any point of S on the boundary of the convex hull of S.

Let N consist of the point O and the 8 points of S which are *nearest* to O. It could happen, of course, that there are any number of points of S which are at the same distance from O. By stipulating the 8 points nearest O, we simply require that the rest of the points of S, that is, the points of $S - N$, all be as far from O as the farthest point of N is away from O.

Since O is on the boundary of the convex hull of S, it will also be on the boundary of the convex hull of N. By part (a) above, then, some point A of N will be at least a distance of $\sqrt{3}$ from O:

$$OA \geq \sqrt{3}.$$

In this case, every point of $S - N$ is at least $\sqrt{3}$ units from O.

Now the induction hypothesis implies that $S - N$ contains a subset T' of k points, each pair of which is at least $\sqrt{3}$ units apart. Because every point of $S - N$ is at least $\sqrt{3}$ units from O, every point of T' will be at least $\sqrt{3}$ units from O. Accordingly, adding O to T' does not spoil its basic property, and the union

$$T = T' \cup O$$

constitutes a subset of $k + 1$ points of S, each pair of which is at least $\sqrt{3}$ units apart.

Since the property in question holds trivially for $k = 1$, we conclude that, for all $k \geq 1$, a set S of $9k$ points, each pair at least 1 unit apart, contains a subset of k points, each pair of which is at least $\sqrt{3}$ units apart.

Our conclusion follows for $9k = 1980$, $k = 220$.

4. An International Competition (Finland) 1980 (p. 44)

Before looking at question #6 in this set, let us first consider the following engaging problem.

> Problem J-21, Olympiad Corner 20, 1980, p. 316. Prove that the first thousand digits after the decimal point in the value of $(6 + \sqrt{35})^{1980}$ are all 9's.

Whenever $(a + \sqrt{b})^n$ or $(a - \sqrt{b})^n$ is encountered, it is a good idea to see whether the conjugate can be of any help. Very often it is useful to add or subtract these quantities, and so we might well think of considering the sum

$$S = (6 + \sqrt{35})^{1980} + (6 - \sqrt{35})^{1980}.$$

The only difference in the binomial expansions of these conjugates is that every second term in $(6-\sqrt{35})^{1980}$ is negative. Consequently, every second term in their sum gets cancelled and the other terms double up. Hence

$$S = 2\left[6^{1980} + \binom{1980}{2}\cdot 6^{1978}\cdot 35 + \binom{1980}{4}\cdot 6^{1976}\cdot 35^2 + \cdots\right],$$

an *even integer* $2k$. Thus

$$(6+\sqrt{35})^{1980} = 2k - (6-\sqrt{35})^{1980}.$$

However, $6 - \sqrt{35} = .0839\ldots$, and we have

$$0 < 6 - \sqrt{35} < .1,$$
$$0 < (6 - \sqrt{35})^{1980} < 10^{-1980}.$$

That is, $(6 + \sqrt{35})^{1980}$ is short of being the integer $2k$ by an amount that is less than $.000\ldots001$, containing 1979 0's. Thus not only the first thousand digits, but almost the first two thousand digits of $(6+\sqrt{35})^{1980}$ are all 9's.

Now let's try problem #6 in the Finnish Competition.

6. Determine the digits *on either side* of the decimal point in the value of $(\sqrt{2} + \sqrt{3})^{1980}$.

First of all we can simplify the expression to

$$(\sqrt{2} + \sqrt{3})^{1980} = [(\sqrt{2} + \sqrt{3})^2]^{990} = (5 + 2\sqrt{6})^{990}$$
$$= (5 + \sqrt{24})^{990}.$$

Proceeding as above, we again see that almost the first *thousand* digits *after* the decimal point are all 9's ($5 - \sqrt{24}$ is quite near .1). Now the number $2k$ is an integer, and if we could determine its last digit d, then the digit *before* the decimal point in $(5 + \sqrt{24})^{1980}$ would be just $d - 1$. The last digit of an integer, of course, is its remainder when divided by

10. Accordingly, we have modulo 10

$$S = (5 + \sqrt{24})^{990} + (5 - \sqrt{24})^{990}$$

$$= 2\left[5^{990} + \binom{990}{2} \cdot 5^{988} \cdot 24 + \binom{990}{4} \cdot 5^{986} \cdot 24^2\right.$$

$$\left. + \cdots + \binom{990}{988} \cdot 5^2 \cdot 24^{494} + \binom{990}{990} \cdot 24^{495}\right]$$

$$\equiv 2 \cdot \binom{990}{990} \cdot 24^{495} \equiv 2 \cdot 24 \cdot (24^2)^{247} \equiv 48 \cdot 576^{247}$$

$$\equiv 8 \cdot 6 \equiv 8$$

(all powers of 576 also end in 6), implying that the digit before the decimal point in $(\sqrt{2} + \sqrt{3})^{1980}$ is a 7.

5. Romanian Olympiad, 1978, 10th Class, Final Round (p. 46, top)

4. A round-robin chess tournament has n contestants, each of whom plays at most once in a round. Determine the minimum number of rounds $f(n)$ necessary to conduct the tournament.

One formula for the answer is

$$f(n) = \left[n - \frac{(-1)^n}{2}\right],$$

where $[x]$ denotes the integer part of x, that is, the greatest integer $\leq x$. All that this really says is

$$f(n) = n \quad \text{for } n \text{ odd, and} \quad f(n) = n - 1 \quad \text{for } n \text{ even.}$$

n	3	4	5	6	7	8	9	$\cdots$
$F(n)$	3	3	5	5	7	7	9	$\cdots$

If n is odd, it is impossible to pair up all n contestants in a round, implying that in each round at least one person must get a bye. Thus n rounds is an absolute minimum for n odd:

(i) a contestant must play each of the other $n - 1$ people, and
(ii) if he were to sit out for a round, a total of at least n rounds would have to be held.

On the other hand, suppose that n rounds were to suffice for n odd. In this case, each contestant would be forced to sit out exactly one round. By the pigeonhole principle, if some round were to have more than one person sitting out, some other round would have to have nobody sitting out, which is impossible for n odd. Thus, if n rounds suffice for n odd, each contestant would get exactly one bye and there would be exactly one bye in each round. In view of this, a new contestant could be fully accommodated without holding any extra rounds by simply having him play the person with the bye in each round; in this way he gets to play everybody else, and a larger tournament with $n + 1$ players is obtained. That is to say, if $f(n) = n$ for n odd, there would also exist a tournament for the even number of players $n + 1$ which also requires only n rounds. But this is again an absolute minimum, for each contestant must play each of the others.

Hence, if $f(n) = n$ for n odd, then for n even we would have

$$f(n) = n - 1.$$

Accordingly, we need only prove that $f(n) = n$ whenever n is odd in order to justify the formula given at the beginning. We shall do this by actually showing how to construct a schedule for $2k + 1$ players $1, 2, \ldots,$ $2k + 1$.

To this end, let contestants be represented by the vertices of a regular $(2k + 1)$-gon $P = P_1 P_2 \ldots P_{2k+1}$. Using vectors, it is easy to see that in the case of a regular polygon having an *odd* number of sides, each side is parallel to a set of $k - 1$ diagonals. Thus a side $P_i P_j$ and the $k - 1$ diagonals parallel to it provide a set of k pairs of subscripts, $(i, j), (i - 1, j + 1), (i - 2, j + 2), \ldots,$ which we can copy directly into a round of the tournament, the unpaired vertex opposite side $P_i P_j$ designating the player to be awarded the bye in that round. The $2k + 1$ rounds thus generated constitute the whole tournament.

It is not difficult to see that every possible match is accounted for once and only once in this scheme. A match is represented exactly once in the polygon P either by a side or a diagonal. But in a regular polygon with an *odd* number of sides, no two sides are parallel and every diagonal is parallel to precisely one side.

The case of 7 players (and thus of 8 players, too) is shown below.

ROUND	SIDE	MATCHES	BYE
1	P_4P_5	$(2,7),(3,6),(4,5)$	1
2	P_5P_6	$(1,3),(4,7),(5,6)$	2
3	P_6P_7	$(2,4),(1,5),(6,7)$	3
4	P_1P_7	$(3,5),(2,6),(1,7)$	4
5	P_1P_2	$(4,6),(3,7),(1,2)$	5
6	P_2P_3	$(5,7),(4,1),(3,2)$	6
7	P_3P_4	$(1,6),(2,5),(3,4)$	7

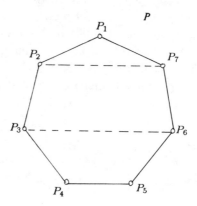

FIGURE 37

6. Romanian Olympiad, 1978, 11th Class, Final Round (p. 46, bottom)

3.   *A* and *B* take turns, *A* first, entering real numbers in an empty 3×3 determinant D until D is completely determined (thus *A* gets 5 turns and *B* only 4). Prove that either player can force D to wind up with the value zero despite the other player's efforts to prevent it.

Let the places in D be labeled

$$D = \begin{vmatrix} a & b & c \\ d & e & f \\ g & h & i \end{vmatrix} = a(ei - hf) - b(di - gf) + c(dh - ge).$$

(a) *A strategy for A.*   In his first two moves A can set two of a, b, c equal to 0. If B doesn't respond with a nonzero value in the remaining place among a, b, c, then A can make all three of them 0, assuring $D = 0$.

Suppose, then, that after three moves A has set a and c equal to 0 and B has entered some nonzero value for b. At this point, then, we have

$$D = \begin{vmatrix} 0 & b & 0 \\ - & - & - \\ - & - & - \end{vmatrix} = -b(di - gf),$$

and A's only chance is to make $di = gf$. But this is easy: essentially all he needs to do is mimic B's moves the rest of the way.

Whatever B specifies in either pair $(d, i), (g, f)$, A specifies in the other pair. For example, if B were to play $g = x$, then A could reply $d = x$; then if B continued with $i = y$, A would clinch things with $f = y$. If B should temporize by playing one of the two empty places e or h, then A also marks time by playing in the other of these two. B cannot avoid having to commit himself to opening up play in the pairs $(d, i), (g, f)$ before A has to, thus permitting A to build up $di = gf$ with his mimicing ways.

(b) *A strategy for B.*   With only 4 of the 9 moves, it is undoubtedly a greater challenge for B to prescribe the value of D. However, we shall see that 4 well chosen 0's are always enough to achieve $D = 0$.

If B were able to make the 4 entries in any 2×2 minor all 0's, then two rows of D would be proportional and $D = 0$;

$$\begin{vmatrix} - & - & - \\ - & 0 & 0 \\ - & 0 & 0 \end{vmatrix}, \quad \begin{vmatrix} - & - & - \\ 0 & - & 0 \\ 0 & - & 0 \end{vmatrix}, \quad \begin{vmatrix} 0 & - & 0 \\ - & - & - \\ 0 & - & 0 \end{vmatrix}, \dots$$

Now, no matter where A begins play, there remain lots of empty 2×2 minors that B might try to fill with 0's; in fact, because of the symmetry in the expansion for D, all of A's first moves are equivalent. If B is able to handle things when A's first move is to specify a, then an appropriate transformation will provide the same solution for any other first move A might make. Let us assume, then, that A begins play at position a.

Obviously B must have more options to exercise than A is able to spoil. While there are many more than 3 empty 2×2 minors after A's first move, there are a particular 3 which overlap in a single common place x; in fact, there is always more than one such set. In any case, if B picks such a set and plays $x = 0$, he keeps open 3 options which no longer have a place in common that can be used by A to spoil them all at once (see the matrices below).

$$
\begin{vmatrix} a & - & - \\ - & x & - \\ - & - & - \end{vmatrix}, \quad
\begin{vmatrix} a & - & - \\ - & x & - \\ - & - & - \end{vmatrix}, \quad
\begin{vmatrix} a & - & - \\ - & x & - \\ - & - & - \end{vmatrix}, \Rightarrow
\begin{vmatrix} a & - & - \\ - & 0 & - \\ - & - & - \end{vmatrix}.
$$

A's second move could be any of the three equivalent pairs (b or d), (c or g), (f or h), or at i. At each succeeding stage, however, B can keep A busy stopping the completion of a whole line of 0's while he proceeds himself to complete his 2×2 minor of all 0's; the figures below show his moves in the order $1, 2, 3$. In each case, A is forced to respond at u and v in a losing cause.

	b (or d)			c (or g)			f (or h)			i		
A's move:	a	b	v	a	u	c	a	u		a	u	
	u	0	1		0	3	3	0	f	2	0	v
		3	2	v	1	2	2	1	v	3	1	i

Key moves for B are given below for opening moves of b and e by A.

$$
\begin{vmatrix} - & b & - \\ x & - & - \\ - & - & - \end{vmatrix}, \quad
\begin{vmatrix} x & - & - \\ - & e & - \\ - & - & - \end{vmatrix}.
$$

7. Moscow Olympiad, 1979, a grade 7 problem (p. 72)

1 (a). A is a given point in the plane. Is it possible to draw 5 circles, none of which covers A, such that *any* ray which emanates from A makes contact with at least 2 of the circles?

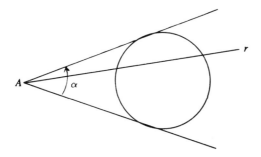

FIGURE 38

Obviously a ray from a point A outside a given circle makes contact with the circle if and only if it lies in the angle α between the tangents from A. In order to contact 2 circles, then, a ray would have to lie in 2 such angles α; conversely, for 2 circles to enlist a given ray r among their contacts, each of their angles α would have to *cover* r. To ensure such a ray of double contact in *every* direction from A, the angles $\alpha_1, \alpha_2, \ldots$ of a set of circles would have to double cover the entire plane around A. The sum of all the angles α, then, would have to amount to at least 2 revolutions, i.e., 720 degrees and, in particular, 5 such angles would have to average at least $720/5 = 144$ degrees.

If any 5 angles $\alpha_1, \alpha_2, \ldots, \alpha_5$ from the range $[144°, 180°)$ are laid arm-to-arm around A, they provide at least a 2-layer coverage of the plane, and 5 angles, each equal to 144 degrees, will do it exactly. Clearly, our requirements are met by any set of 5 circles, each tangent to the arms of such an angle α, each to a different α (see the figures.)

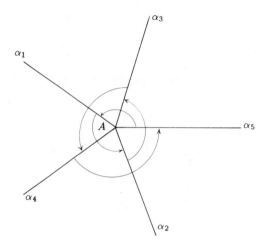

FIGURE 39a

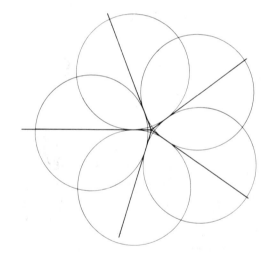

FIGURE 39b

8. Moscow Olympiad, 1979, a grade 10 problem (p. 74)

10. S is a collection of disjoint intervals in the unit interval
$[0, 1]$. If no two points of S are 1/10th of a unit apart, prove

that the sum of the lengths of all the intervals in S cannot exceed $1/2$.

Since a span of $1/10$th of a unit is obviously an important distance in connection with the set S, one might well think of partitioning the unit interval into 10 equal parts as in the decimal system:

$$0 \quad .1 \quad .2 \quad .3 \quad .4 \quad .5 \quad .6 \quad .7 \quad .8 \quad .9 \quad 1$$

Since no two points of S are $1/10$th of a unit apart, we see that if the part of S between 0 and $.1$ were to be shifted $1/10$th of a unit to the right to occupy corresponding places between $.1$ and $.2$, there would be no overlapping of points of S, and we conclude that all the points of S between 0 and $.2$ can be accommodated without overlapping in the upper half of this range. Because the same is true for each of the other 4 ranges—$.2$ to $.4$, $.4$ to $.6$, $.6$ to $.8$, $.8$ to 1—the 5 upper halves of these 5 ranges suffice for the entire set S, implying that the "measure" of S cannot exceed $5(1/10) = 1/2$, as required.

9. The Second Selection Test for the Romanian International Olympiad Team, 1978 (p. 76)

4. parts (a), (c), and (d). M is a set of $3n$ points in the plane such that the maximum distance between any two of the points is 1 unit. Prove that

(a)   for any 4 points of M, the distance between some two of them is less than or at most $1/\sqrt{2}$,

(c)   some circle of radius $\leq \sqrt{3}/2$ encloses the entire set M,

(d)   there is some pair of the $3n$ points of M whose distance apart is at most $4/(3\sqrt{n} - \sqrt{3})$.

(a)   There are only two ways a set of 4 points can occur in the plane: either

(i) they determine a convex quadrilateral (including the degenerate case of collinear points), or

(ii) one point occurs inside or on the triangle formed by the other three.

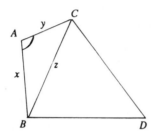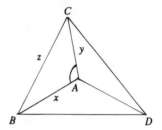

FIGURE 40

Since the 4 angles of a convex quadrilateral add up to 360°, they can't all be less than a right angle; also in the nonconvex case (ii), one of the 3 angles at the "inner" point must be at least a right angle (they average $360/3 = 120°$).

In any case, some 3 of a subset of 4 points of M must determine a triangle ABC containing an angle which is at least as big as a right angle. In such a triangle ABC, suppose that $A \geq 90°$ and that the sides are

$$AB = x, \quad AC = y, \quad \text{and} \quad BC = z.$$

By the law of cosines we have

$$z^2 = x^2 + y^2 - 2xy \cdot \cos A \geq x^2 + y^2,$$

since $A \geq 90°$ makes $\cos A \leq 0$. Thus, if both x and y were to exceed $1/\sqrt{2}$, we would have

$$z^2 \geq x^2 + y^2 > \frac{1}{2} + \frac{1}{2} = 1,$$

giving the contradiction $z > 1$ (no 2 points of M are farther apart than 1). Hence at least one of x or y must be $\leq 1/\sqrt{2}$.

(c)   Part (c) is quite easy and is undoubtedly included in the problem as a helpful preliminary to the much more difficult part (d).

Since a maximum is always attained (which is not necessarily true for a least upper bound), there must be some two points A and B of M which actually are at the maximum distance of 1 unit apart. Since no 2 points of M are farther apart than 1 unit, a circle of radius 1, drawn about any point of M as center, must enclose the entire set M. In particular, each of the unit circles $A(1)$ and $B(1)$ must contain the whole set, implying that M must reside in their lens of intersection $AXBY$.

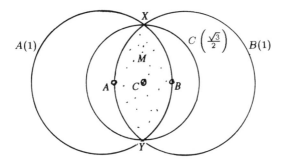

FIGURE **41**

Because $\triangle ABX$ is an equilateral triangle of unit side, the midpoint C of AB is a distance of $\sqrt{3}/2$ from X (XC is an altitude). Clearly, the circle $C(\sqrt{3}/2)$ covers the convex lens $AXBY$ and thus the set M.

(d)   Since 2 points in a circle cannot be farther apart than the diameter, we would be able to conclude that some 2 points of M are no farther apart than $4/(3\sqrt{n} - \sqrt{3})$ if we could show that they both belonged to some circle of radius $r = 2/(3\sqrt{n} - \sqrt{3})$.

From this point of view, the problem amounts to showing that, no matter how the points of M may be scattered about, it is always possible to place a circle of radius r ($= 2/(3\sqrt{n} - \sqrt{3})$) on the plane so as to cover at least 2 of them. Accordingly, the pigeonhole principle technique used in Morsel 17 (The Circle and the Annulus) comes to mind as a possible line of attack.

To this end, let a circle of the above radius r be drawn about *each* point of M as center. For points of M that are near the boundary of the enclosing circle $C(\sqrt{3}/2)$, these little circles stick out past the circumference of $C(\sqrt{3}/2)$. However, since their centers are all in $C(\sqrt{3}/2)$, this entire set S of little circles would be contained in the larger concentric circle $C(\sqrt{3}/2 + r)$.

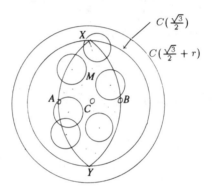

FIGURE 42

Now, if we can show that the areas of all the little circles in S add up to *more* than the area of the enclosing circle $C(\sqrt{3}/2 + r)$, it would follow that some point Z of $C(\sqrt{3}/2 + r)$ must lie buried under at least 2 of the circles $P(r)$ and $Q(r)$, $P, Q \in M$ (this follows from the continuous version of the pigeonhole principle: if no point of the underlying circle were covered more than once, then the total area of the covers couldn't add up to more than that of the underlying circle).

Of course, the bigger the circle, the more difficult it is for the little covering circles to meet the demand. And clearly, the enclosing circle $C(\sqrt{3}/2 + r)$ introduces an underlying region to be covered by S that is larger than it need be. A margin of width r around the lens $AXBY$ would have sufficed, and would make a considerably lighter demand on the total area of S; also, it is the obvious choice for this role. However, although we impose a more stringent requirement on the set S by using the larger circle, the computational benefits that accrue from dealing with a *circle* are overwhelming. Without that helpful hint in part (c), we probably would not have thought of backing off from the lens to the circle and would soon have become mired in an algebraic tangle in trying to deal with the lens and its margin. We can only hope that this move does not give away more than the method has slack to give.

Proceeding, we note that the circle's area is $\pi(\sqrt{3}/2 + r)^2$. Since each of the $3n$ points of M is the center of a circle of radius r, the total area of the circles in S is $3n\pi r^2$. If we can show that

$$3n\pi r^2 > \pi \left(\frac{\sqrt{3}}{2} + r \right)^2 ,$$

then we would have established the existence of two points P and Q of M whose circles $P(r)$ and $Q(r)$ overlap in some point Z. In this case, the distances PZ and QZ would each be $\leq r$, and therefore the circle $Z(r)$ would cover both the points P and Q, giving

$$PQ \leq 2r = \frac{4}{3\sqrt{n} - \sqrt{3}}, \qquad \text{as desired.}$$

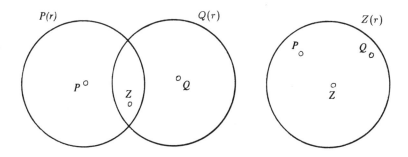

FIGURE 43

It remains to show, for $n \geq 1$, that

$$3n\pi r^2 > \pi \left(\frac{\sqrt{3}}{2} + r \right)^2 ,$$

that is,

$$3nr^2 > r^2 + \sqrt{3} \cdot r + \frac{3}{4},$$

$$(3n-1)r^2 - \sqrt{3} \cdot r - \frac{3}{4} > 0,$$

$$4(3n-1)r^2 - 4\sqrt{3} \cdot r - 3 > 0,$$

$$r[4(3n-1)r - 4\sqrt{3}] - 3 > 0.$$

Substituting for r, this requires

$$\frac{2}{3\sqrt{n} - \sqrt{3}} \left[\frac{4(3n-1) \cdot 2}{3\sqrt{n} - \sqrt{3}} - 4\sqrt{3} \right] - 3 > 0,$$

which easily simplifies to $21n - 1 - 6\sqrt{3} \cdot \sqrt{n} > 0$.

Thus we need to show that the function

$$21(\sqrt{n})^2 - 6\sqrt{3}(\sqrt{n}) - 1 \qquad \text{is positive for all } n \geq 1.$$

Now, the positive root of the equation $21(\sqrt{n})^2 - 6\sqrt{3}(\sqrt{n}) - 1 = 0$ is

$$\sqrt{n} = \frac{6\sqrt{3} + \sqrt{108 + 84}}{42} < \frac{12 + 14}{42} < 1.$$

Hence the function *is* positive for $\sqrt{n} \geq 1$, and the proof is complete.

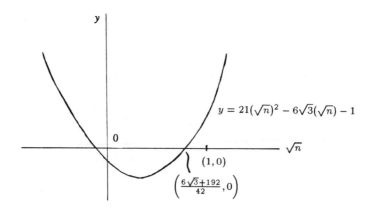

FIGURE 44

10. Third Selection Test for the Romanian International Olympiad Team, 1978 (p. 77)

6.　Prove that the expression $x\sqrt{2} + y\sqrt{3} + z\sqrt{5}$, where x, y, and z are *integers* which are *not all zero*, can be made arbitrarily close to zero.

First of all, we note that $\sqrt{2}, \sqrt{3}, \sqrt{5}$ are linearly independent: i.e.,

$$x\sqrt{2} + y\sqrt{3} + z\sqrt{5} = 0, \quad x, y, z \text{ integers} \Rightarrow x = y = z = 0.$$

Although the easy proof of this fundamental property was probably not required of the olympiad hopefuls who tried this test, we include it here for completeness.

　Suppose $x\sqrt{2} + y\sqrt{3} + z\sqrt{5} = 0$, where x, y, z are integers. Then $z\sqrt{5} = -(x\sqrt{2} + y\sqrt{3})$, and squaring gives $5z^2 = 2x^2 + 2xy\sqrt{6} + 3y^2$, from which we obtain the impossible result that $\sqrt{6}$ is *rational* unless $xy = 0$. Hence at least one of x and y must be zero.

　Thus we have either $y\sqrt{3} + z\sqrt{5} = 0$ or $x\sqrt{2} + z\sqrt{5} = 0$, and squaring in either case gives a similar impossible rational value for either $\sqrt{15}$ or $\sqrt{10}$ unless another of x, y, z is also zero. This makes 2 of them equal to zero, and the initial assumption forces the third to be the same.

　Let the set of numbers in question be denoted by S:

$$S = \{x\sqrt{2} + y\sqrt{3} + z\sqrt{5} \mid x, y, z \in Z, x^2 + y^2 + z^2 \neq 0\}.$$

Since x, y, z are not all zero, the linear independence of $\sqrt{2}, \sqrt{3}, \sqrt{5}$ implies that 0 does *not* belong to S.

　Now let's generate a sequence of numbers $d_1 > d_2 > d_3 > \cdots$ that *do* belong to S. To begin, for $(x, y, z) = (-1, 1, 0)$, let

$$d_1 = \sqrt{3} - \sqrt{2} = .31\ldots \ .$$

Next, let d_2 be the remainder when $\sqrt{5}$ is divided by d_1:

$$\sqrt{5} = q_1 d_1 + d_2,$$

q_1 is an integer (actually 7) and $0 \le d_2 < d_1$.

Since

$$d_2 = \sqrt{5} - q_1 d_1 = \sqrt{5} - 7\sqrt{3} + 7\sqrt{2},$$

we have $d_2 \in S$. Because 0 does not belong to S, then d_2 cannot be 0. Accordingly, we can divide d_1 by d_2 to obtain another remainder d_3. In general, let succeeding d's be the (nonnegative) remainder when d_{i-1} is divided by d_i:

$$d_{i-1} = q_i d_i + d_{i+1}, \qquad q_i \text{ an integer and } 0 \le d_{i+1} < d_i.$$

Clearly, as we saw above, if d_{i-1} and d_i both belong to S, then, because q_i is an integer, d_{i+1} will also belong to S. In this case, the remainder can't be 0, and we have

$$0 < d_{i+1} < d_i \qquad \text{for all } i.$$

Clearly, S contains an unending decreasing sequence of positive real numbers

$$D = \{d_1 > d_2 > d_3 > \cdots > d_i > d_{i+1} > d_{i+2} > \cdots\}.$$

However, a decreasing sequence of positive numbers doesn't necessarily get arbitrarily small. Let us now show that this *is* true of D.

It is easy to see that no d_{i+1} could ever be exactly $\frac{1}{2} d_i$ for, in that case, in the next division, to get d_{i+2}, the quotient would be 2 and the remainder $d_{i+2} = 0$:

$$d_i = 2 \cdot d_{i+1} + 0.$$

Thus, succeeding terms d_{i+1} are always either greater than or less than $\frac{1}{2} d_i$. Now if D were to drop by a factor of $\frac{1}{2}$ at each succeeding term, its terms would certainly get arbitrarily small. Consider the situation when this doesn't happen, that is, when

$$d_{i+1} > \frac{1}{2} d_i.$$

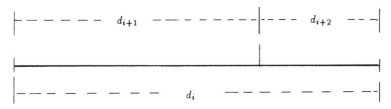

FIGURE 45

$$d_{i+1} > \frac{1}{2}d_i \Rightarrow d_i = 1 \cdot d_{i+1} + d_{i+2}, \text{ where } d_{i+2} < \frac{1}{2}d_i.$$

In this case, the quotient in the next division would be just 1, and the remainder d_{i+2}, however it might compare with its own divisor d_{i+1}, would certainly be less than $\frac{1}{2}d_i$.

That is to say, if d_{i+1} isn't less than $\frac{1}{2}d_i$, then d_{i+2} certainly *is*. Consequently, in all cases the sequence D drops by a factor of more than $1/2$ *every one or two terms*, and thus gets arbitrarily close to zero.

11. Canadian Olympiad, 1981 (p. 140)

3.   No matter *how many* straight lines there may be in a finite set S, or how they may *crisscross* the plane, prove that there are arbitrarily large circles in the plane which do not meet any of them; on the other hand, prove that there is a denumerably infinite set of straight lines L such that *every circle* in the plane, no matter how small, meets at least one of the lines in L.

Once the lines in S have been chosen, any points of intersection they may produce are determined. From any point O in the plane, a perpendicular can be drawn to each line in S, and thus a circle $O(r)$, where r is greater than any of these perpendiculars, will cross each line in S. Expanding if necessary, this circle can also be made large enough to encompass all the points of intersection determined by S. Thus, let C be a circle, with an arbitrary point O as center, that both crosses every line in S and contains all its points of intersection (see the figures).

The lines of S emerge from C as rays which partition its infinite exterior. Since all the points of intersection are contained in C, no two of these rays can cross. Consequently, either

(i) all the rays are parallel, or

(ii) some *adjacent* pair of rays *diverge*.

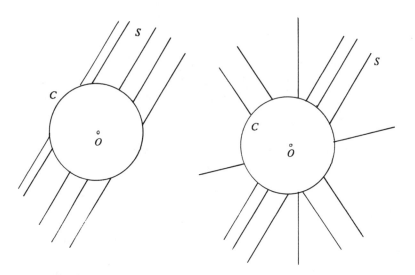

FIGURE 46

In case (i), there is a blank half plane on each side of C in which arbitrarily large circles can be drawn; and in case (ii), no matter how slowly a region may diverge, it will eventually become wide enough to accommodate in its interior a circle of any size.

On the other hand, because the rational numbers can be enumerated, the set L of straight lines $y = mx$, with m rational, is denumerably infinite. Now, any circle which contains the origin O in its interior crosses every line in L. The tangents from O to any other circle determine between them a divergent region R (a half plane if the circle actually goes through O). (See Figure 47.) No matter how slowly R may diverge, the span d across R along the lattice lines $x = a$, a a nonzero integer, must eventually exceed 1 unit, implying that R cannot avoid

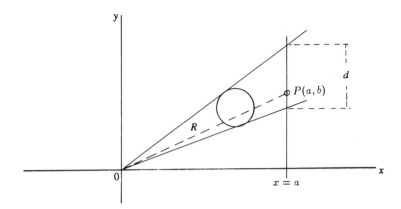

FIGURE 47

containing some lattice point $P(a, b)$. Thus the circle crosses the line $y = mx$, where $m = b/a$ is rational, making OP a member of L.

12. U.S.A. Olympiad, 1981 (p. 141)

1. If a given angle θ is $1/n$th of a straight angle, where n is a given positive integer that is not a multiple of 3, prove that θ can be trisected with just straightedge and compasses.

Since n is given, let the quotient k and remainder r be calculated when n is divided by 3:

$$n = 3k + r.$$

Because n is not a multiple of 3, the value of r will be either 1 or 2.

Now, one angle that we *are* able to construct with Euclidean tools is a 60° angle. From the given fact

$$\theta = \frac{180}{n} \qquad \text{degrees,}$$

we have

$$60 = \frac{n}{3}\theta = \left(k + \frac{r}{3}\right)\theta = k\theta + r \cdot \frac{\theta}{3},$$

giving

$$r \cdot \frac{\theta}{3} = 60 - k\theta.$$

Since k is known, the angle $60 - k\theta$ can be constructed. If $r = 1$, the result is the desired $\theta/3$; if $r = 2$, a simple bisection is required.

13. An Unused Problem from the International Olympiad, 1981 (p. 237)

> 12.   Consider a cluster of n spherical planets, all the same size, which drift rigidly together through outer space, that is, each with no motion relative to the others in the cluster. In general, there is some region R on the surface of a planet which cannot be seen from any point on any of the other planets (if a planet is in the midst of the others, then its surface is completely visible and this region is empty). Prove that the sum of the areas of these invisible regions R is exactly equal to the area of the surface of one planet.

Solution by Allen Schwenk (Western Michigan University). In the finest problem-solving tradition, Professor Schwenk obtains the key to this problem from the analogous 2-dimensional problem. If a circular planet lies completely within the convex hull H of a similarly defined system of circular planets, its entire circumference is visible from the other planets. In fact, it is evident that the invisible arcs R of the various circles are precisely the circular parts, $a, b, c, \ldots$, of the boundary of H, that is, the arcs of intersection of the planets and their convex hull H. (See Figure 48.)

Let these invisible arcs R be transferred to a score-keeping circle S of the same size as each of the planets. This can be done very neatly by letting a tangent t roll around H at the same time as a tangent T rolls around S, the two tangents always having the same direction and occurring, respectively, on the same sides of H and S. Clearly, as the point of contact p of t traverses an arc a, the corresponding point of contact P of T traces out an identical arc on the circumference of S. As t rolls all the way around H, T rolls around S, reproducing consecutively on S all the invisible arcs $a, b, c, \ldots$, the point P marking a boundary

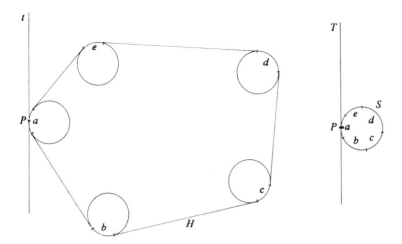

FIGURE 48

point between consecutive arcs when t becomes a common tangent to two or more of the circles. The conclusion follows.

Similarly, in the 3-dimensional problem, tangent *planes* t and T rolling around the convex hull H and an equal score-keeping sphere S copy all the invisible regions R onto S, thus showing not only that the sum of the areas of the regions R is equal to the surface of one planet, but further that they actually *fit together to cover* a planet exactly.

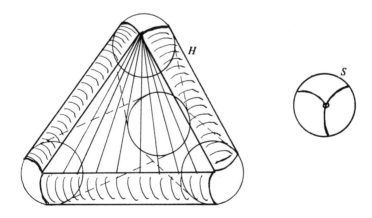

FIGURE 49

14. Hungarian Olympiad, second round, 1981 (p. 267)

1. (first version).   For which positive integers n is
$N = 2^8 + 2^{11} + 2^n$ a perfect square?

We have $N = 256 + 2048 + 2^n = 2304 + 2^n = 48^2 + 2^n$. Since
$49^2 = 2401$ and $50^2 = 2500$, it is easy to check that no n less than 8
makes N a perfect square.

Suppose that $n \geq 8$ and that

$$N = 2^8 + 2^{11} + 2^n = k^2 \qquad \text{for some integer } k.$$

Then

$$2^8(1 + 8 + 2^{n-8}) = k^2,$$

where $1 + 8 + 2^{n-8}$ is a positive integer q. Since 2^8 and k^2 are both
squares, q must also be a perfect square:

$$q = 9 + 2^{n-8} = t^2 \qquad \text{for some positive integer } t.$$

Hence

$$t^2 - 9 = 2^{n-8},$$

$$(t-3)(t+3) = 2^{n-8},$$

implying that $t - 3$ and $t + 3$ are powers of 2 that differ by 6. Thus $t - 3$
and $t + 3$ must be 2 and 8, making $t = 5$.

Then

$$2^{n-8} = t^2 - 9 = 25 - 9 = 16 \Rightarrow n - 8 = 4 \quad \text{and} \quad n = 12.$$

Hence $n = 12$ is the only positive integer n that makes N a perfect
square.

3. (second version).   If $f(n)$ denotes the number of 0's in
the decimal representation of the positive integer n, what is
the value of the sum

$$S = 2^{f(1)} + 2^{f(2)} + \ldots + 2^{f(9999999999)}?$$

Consider first the set of positive integers which has r digits. The number of 0's in such an integer could be anything from 0 to $r - 1$ (the leading digit can't be 0). Let T be the subset which has k 0's, that is, the subset of r-digit positive integers which has exactly k 0's.

For each integer in T we have $f(n) = k$ and $2^{f(n)} = 2^k$. In order to determine the total contribution to S of the numbers in T, we need only find out how many of them there are. Now, to construct an integer in T, we would have to do two things—

(i) select k places for the 0's from the $r - 1$ places following the leading digit and

(ii) select nonzero digits for each of the other $r - k$ places.

Clearly (i) can be done in $\binom{r-1}{k}$ ways and (ii) in 9^{r-k} ways, implying that the number of integers in T is

$$\binom{r-1}{k} \cdot 9^{r-k}.$$

Therefore the total contribution to S of these numbers is

$$\binom{r-1}{k} \cdot 9^{r-k} \cdot 2^k.$$

Adding the contributions for all $k = 0, 1, 2, \ldots, r - 1$, we find that the entire contribution to S of the r-digit numbers is

$$S_r = \sum_{k=0}^{r-1} \binom{r-1}{k} \cdot 9^{r-k} \cdot 2^k = 9^r \sum_{k=0}^{r-1} \binom{r-1}{k} \left(\frac{2}{9}\right)^k$$

$$= 9^r \left(1 + \frac{2}{9}\right)^{r-1}, \qquad \text{by the binomial theorem,}$$

$$= 9(9 + 2)^{r-1} = 9 \cdot 11^{r-1}.$$

Since the sum S contains $2^{f(n)}$ for all positive integers having 10 or fewer digits, we get finally that

$$S = S_1 + S_2 + \cdots + S_{10}$$
$$= 9(1 + 11 + 11^2 + \cdots + 11^9)$$
$$= 9 \cdot \frac{11^{10} - 1}{10} = 23343682140.$$

ON A BALANCED
INCOMPLETE BLOCK DESIGN

In the array below, the 8 integers 0, 1, 2, 3, 4, 5, 6, 7 are laid out in 14 rows, 4 to a row. The arrangement is called a balanced incomplete block design (BIBD) and has two special features:

(a) each integer occurs the same number of times (7),
(b) each pair of integers occurs together in the same row the same number of times (3).

$$
\begin{array}{cccc}
0 & 1 & 2 & 7 \\
1 & 3 & 5 & 7 \\
2 & 4 & 5 & 7 \\
0 & 3 & 4 & 7 \\
1 & 4 & 6 & 7 \\
2 & 3 & 6 & 7 \\
0 & 5 & 6 & 7 \\
3 & 4 & 5 & 6 \\
0 & 2 & 4 & 6 \\
0 & 1 & 3 & 6 \\
1 & 2 & 5 & 6 \\
0 & 2 & 3 & 5 \\
0 & 1 & 4 & 5 \\
1 & 2 & 3 & 4 \\
\end{array}
$$

The rows are called *blocks* and in general a BIBD is described by its parameters (v, b, r, k, λ), where

v = the number of elements,
b = the number of blocks,
r = the number of times each element occurs,
k = the number of elements to a block (a constant),
λ = the number of times each pair will be found
            together in a block.

Thus the design above is an $(8, 14, 7, 4, 3)$.

BIBD's have application in the design of experiments. For example, instead of treating each of 8 test patients with all 14 experimental medications, medication M_i could be given to just the 4 people indicated by block i; this permits a statistical analysis which yields information that, without the regularity of the BIBD, could only be obtained from an experiment of much larger dimensions.

It is easy to see that the parameters of a BIBD are subject to the following two simple conditions:

(a) $bk = vr$ = the total number of elements altogether,

(b) $\lambda(v - 1) = r(k - 1)$ = the total number of pairs containing the element i (i is paired λ times with each of the other $v - 1$ elements, and each of the r blocks that contains i generates $k - 1$ of them with the other $k - 1$ elements in that block).

Finally, our problem:

for the BIBD(v, b, r, k, λ), prove that

$$k^r \geq v^\lambda.$$

Solution

By the relation (b) above, we have

$$\lambda(v - 1) = r(k - 1)$$
$$\lambda v - \lambda = rk - r$$

and

$$\lambda v + (r - \lambda) = rk.$$

That is to say, the set of r positive integers

$$(v, v, \ldots, v, 1, 1, \ldots, 1),$$

consisting of λ v's and $(r - \lambda)$ 1's, has sum rk. Thus, from the well known arithmetic mean-geometric mean inequality $A \geq G$, we have

$$A = \frac{\text{sum}}{r} = \frac{rk}{r} = k \geq G = (v^\lambda \cdot 1^{r-\lambda})^{1/r},$$

and

$$k^r \geq v^\lambda.$$

THE RED AND WHITE BALLS

An urn contains w white balls ($w \geq 3$) and r red balls. If 3 balls were to be withdrawn without replacement, the probability they would all be white is p. An extra white ball in the urn would increase this probability by a third of its value. If r is as great as these conditions allow, how many red balls are there in the urn?

Solution

Among the $\binom{w+r}{3}$ ways of selecting 3 balls from the urn, there are $\binom{w}{3}$ ways of picking 3 white ones, and we have

$$p = \frac{\binom{w}{3}}{\binom{w+r}{3}}. \tag{1}$$

Similarly, with an extra white ball, we have

$$\frac{4}{3}p = \frac{\binom{w+1}{3}}{\binom{w+r+1}{3}}. \tag{2}$$

Substituting in equation (2) the value of p obtained from equation (1), we get

$$\frac{4}{3} \cdot \frac{\frac{w!}{3!(w-3)!}}{\frac{(w+r)!}{3!(w+r-3)!}} = \frac{\frac{(w+1)!}{3!(w-2)!}}{\frac{(w+r+1)!}{3!(w+r-2)!}},$$

$$\frac{4}{3} = \frac{\frac{w+1}{w-2}}{\frac{w+r+1}{w+r-2}} = \frac{(w+1)(w+r-2)}{(w-2)(w+r+1)},$$

$$4(w-2)(r+w+1) = 3(w+1)(r+w-2)$$

$$r(w-11) = -w^2 + w + 2$$

$$r = \frac{-w^2 + w + 2}{w - 11}$$

$$= \frac{-w^2 + 11w - 10w + 110 - 108}{w - 11}$$

$$= -w - 10 + \frac{108}{11 - w}.$$

Now w is a positive integer ≥ 3, and in order to keep r finite and positive, this relation requires $w < 11$. Hence the only feasible values of w are 3, 4, 5, 6, 7, 8, 9,10. But 3, 4, and 6 fail to make $11 - w$ a divisor of 108, thus preventing r from being an integer. Hence w can only be 5,7,8,9, or 10. When $w = 10$, the value of $108/(11 - w)$ is 108, which far overshadows all other contributions to the value of r, and we see that the maximum case is given by

$$r = -10 - 10 + 108 = 88 \text{ red balls.}$$

Exercise. Show that it is not really necessary to state in the problem that the selections are made "without replacement," for the problem is impossible otherwise.

A PRIME NUMBER GENERATOR

Suppose the first n prime numbers $2, 3, 5, \ldots, p_n$, are divided into two groups in any way whatever and the products A and B of the numbers in the groups determined, an empty set being assigned a product of 1. Prove that each of the numbers $A + B$, and $|A - B|$ will also always turn out to be a prime number for every such partition, provided only that it is less than p_{n+1}^2 (and is > 1, of course.) For example:

(i) For $(2, 3, 5)$: $(p_{n+1}^2 = 7^2 = 49)$

$$2 \cdot 3 + 5 = 11, \quad 2 \cdot 5 + 3 = 13,$$

$$2 \cdot 5 - 3 = 7, \quad 3 \cdot 5 + 2 = 17,$$

$$3 \cdot 5 - 2 = 13, \quad 2 \cdot 3 \cdot 5 + 1 = 31, \quad 2 \cdot 3 \cdot 5 - 1 = 29.$$

(ii) For $(2, 3, 5, 7)$: $(p_{n+1}^2 = 11^2 = 121)$

$$5 \cdot 7 + 2 \cdot 3 = 41, \quad 5 \cdot 7 - 2 \cdot 3 = 29, \quad 2 \cdot 3 \cdot 5 + 7 = 37,$$

$$2 \cdot 3 \cdot 5 - 7 = 23, \quad 3 \cdot 7 + 2 \cdot 5 = 31, \quad 3 \cdot 7 - 2 \cdot 5 = 11, \quad \ldots.$$

Solution

For me, the fascination with this procedure seems to lie to a considerable extent in the amusement of watching it actually churn out prime numbers; I'm sure I only half believed it would work until I had seen it performed a few times. However, the proof, so short and sweet, is also a delight in its own right.

Clearly each of the first n prime numbers divides one of A, B and not the other, and so must fail to divide either $A + B$ or $|A - B|$. Any prime divisor of $A + B$ or $|A - B|$, then, is at least as big as p_{n+1}, and if there were more than one of them, the number would amount to at least p_{n+1}^2, exceeding its stipulated limit. For $A + B$ or $|A - B|$ between 1 and p_{n+1}^2, then, it must itself be a prime number p such that

$$p_{n+1} \leq p < p_{n+1}^2.$$

NEUBERG'S THEOREM

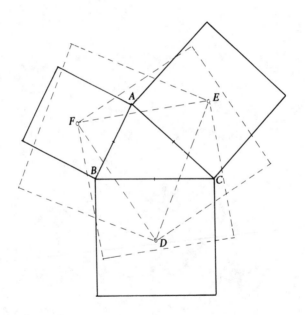

FIGURE 50

If D, E, F are the centers of squares constructed *outwardly* on the sides of a triangle ABC, then the centers of the squares drawn *inwardly* on the sides of triangle DEF are the midpoints of the sides of $\triangle ABC$. (This theorem is due to Joseph Neuberg, 1840–1926.)

Solution

It is sufficient to show that the midpoint M of BC is the center of the square on EF, that is, that MF and ME are equal and perpendicular.

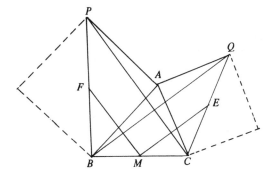

FIGURE 51

Referring to the figure, a quarter-turn about A in the appropriate direction clearly carries $\triangle PAC$ into $\triangle BAQ$, showing that PC and BQ are equal and perpendicular.

But, because F and M are the midpoints of two sides of $\triangle BPC$, FM must be parallel to the third side PC and half as long; similarly EM is parallel to BQ and half its length. Since PC and BQ are equal and perpendicular, then so are FM and EM.

A GEOMETRICAL CALCULATION

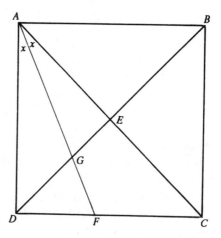

FIGURE 52

The diagonals of a square $ABCD$ meet at E, and the bisector of $\angle CAD$ crosses DE at G and DC at F. If the length of GE is 24, how long is FC?

Solution

In $\triangle ADF$, $\angle DFA = 180 - 90 - 22\frac{1}{2} = 67\frac{1}{2}^\circ$. Now, let a line through E and parallel to CD meet AG at K. Then alternate angles DFK and FKE are equal, making $\angle GKE = 67\frac{1}{2}^\circ$. Similarly, alternate angles FDE and DEK are 45°, making the third angle in $\triangle KEG$, namely EGK, equal to $180 - 45 - 67\frac{1}{2} = 67\frac{1}{2}^\circ$. Thus angles EGK and EKG are $67\frac{1}{2}^\circ$ and $\triangle KEG$ is isosceles with $KE = GE = 24$.

Then, from the obviously similar triangles AKE and AFC we have

$$\frac{FC}{KE} = \frac{AC}{AE} = \frac{2}{1} \implies \frac{FC}{24} = 2,$$

giving $FC = 48$.

ON CUBIC CURVES

P is a point on the graph of $y = x^3$ (see Figure 53). The tangent at P crosses the curve at Q, and A is the area between the curve and the segment PQ. Similarly, the tangent at Q meets the curve again at R, and B is the area between the curve and QR. Prove that B is always 16 times as great as A for every choice of the point P. (Actually $B = 16A$ (it's always 16) for every point P on every curve of the 3rd degree.)

Solution

Let the coordinates of P, Q, and R be $P(k, k^3)$, $Q(t, t^3)$, and $R(s, s^3)$. Then the equation of the tangent PQ is

$$y - k^3 = 3k^2(x - k), \qquad \text{or} \qquad y = 3k^2 x - 2k^3.$$

Solving with $y = x^3$ for the coordinates of P and Q, we get

$$x^3 = 3k^2 x - 2k^3,$$
$$x^3 - 3k^2 x + 2k^3 = 0.$$

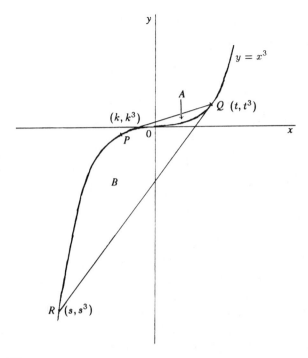

FIGURE 53

Since there is no term in x^2, the sum of the roots of this equation must be zero. But, clearly, these roots are k, k, and t, and so we obtain

$$t = -2k.$$

Similarly,

$$s = -2t.$$

The area A, then, is given by

$$A = \left| \int_k^{-2k} \left[(3k^2 x - 2k^3) - x^3 \right] \, dx \right|$$

$$= \left| \left[\frac{3k^2 x^2}{2} - 2k^3 x - \frac{x^4}{4} \right] \Big|_k^{-2k} \right|,$$

which we can see, without going any farther, is going to reduce to some number of k^4's, and we have

$$A = ck^4, \qquad \text{where } c \text{ is some positive constant.}$$

Now, the only difference that occurs in determining B is that k is replaced by t. Hence

$$B = ct^4, \qquad \text{where } c \text{ is the } same \text{ constant as above.}$$

Since $t = -2k$, then, we have

$$B = c(-2k)^4 = 16(ck^4) = 16A.$$

In the light of this result, my friend and colleague Ian McGee soon came up with the following properties of quartic polynomials. In each figure, P and Q are the points of inflection.

If PQ is extended to cross the curve again at R and S, three regions A, B, and C are caught between the curve and the line (see Figure 54); for all quartic curves, it turns out that the areas of A and C are the same and B is twice as big:

$$A = C \qquad \text{and} \qquad B = 2A.$$

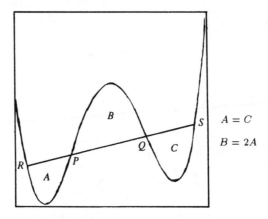

FIGURE 54

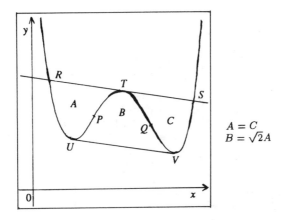

FIGURE 55

Let UV be the double tangent (at both U and V) and let T be the point on the curve which is "directly above" the midpoint of UV, that is, whose x-coordinate is halfway between the x-coordinates of U and V. Then UV and the tangent at T trap three regions $A, B,$ and C against the curve, as shown in Figure 55. In all cases, A and C have equal areas and B is $\sqrt{2}$ times as great. Incidentally, UV, PQ, and RS are parallel.

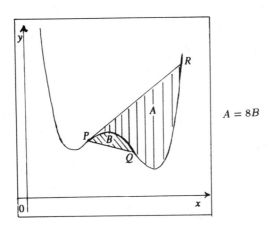

FIGURE 56

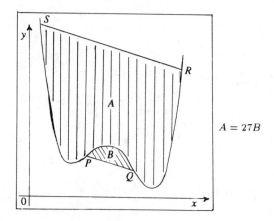

FIGURE 57

PQ and the tangent at P determine disjoint regions A and B, as shown in Figure 56, such that A is always 8 times as big as B.

If both the tangents PR and QS are drawn, then RS and PQ determine disjoint regions A and B such that A is 27 times B (see Figure 57). Also, RS is parallel to PQ.

Professor McGee has also looked briefly at functions of higher degree, but we shall leave the interested reader to investigate these matters on his own.

AN OLYMPIAD PRACTICE PROBLEM

Suppose x and y vary over the nonnegative real numbers. If the value of

$$x + y + \sqrt{2x^2 + 2xy + 3y^2}$$

is always 4, prove that x^2y is always less than 4.

(This problem is adapted from a problem used at a practice session of the U.S.A. Olympiad team, 1977.)

Solution

Even if you were told to use the well known arithmetic mean-geometric mean inequality (and I'm sure that no such hint would be given to the olympiad team members), your problems would be far from over. What good is

$$\frac{x + y}{2} \geq \sqrt{xy} \quad ?$$

If the right side were to contain the pertinent quantity x^2y, then it might amount to something. But is there any way to arrange this ? Since x^2y is the product of three factors, $x, x,$ and y, it might occur to us to try

$$\frac{x + x + y}{3} \geq \sqrt[3]{x^2y}.$$

Now the right side is good, but the left side no longer contains just $x+y$, which is the rational first part of the awkward constraint that governs x and y. Is it possible to get just $x + y$ on the left and $x^2 y$ on the right?; perhaps we can afford the reasonable price of a few constants of adjustment.

If $x/2$ were used twice, instead of x, we would get

$$\frac{\frac{x}{2} + \frac{x}{2} + y}{3} \geq \sqrt[3]{\frac{x}{2} \cdot \frac{x}{2} \cdot y}$$

and

$$x + y \geq 3 \left[\frac{x^2 y}{4} \right]^{1/3},$$

which is very promising.

Encouraged, let's try to get a similar result with $2x^2 + 2xy + 3y^2$ on the left. After a little (?) experimenting, one may eventually come up with the fifteen terms

$$\frac{\left[\frac{2x^2}{8} + \frac{2x^2}{8} + \cdots + \frac{2x^2}{8} \right] + \left[\frac{2xy}{4} + \cdots + \frac{2xy}{4} \right] + \left[y^2 + y^2 + y^2 \right]}{15}$$

(the brackets containing, respectively, 8,4, and 3 like terms)

$$\geq \left[\left(\frac{2x^2}{8} \right)^8 \left(\frac{2xy}{4} \right)^4 (y^2)^3 \right]^{1/15}$$

$$= \left[\frac{x^{20} y^{10}}{2^{20}} \right]^{1/15},$$

that is,

$$2x^2 + 2xy + 3y^2 \geq 15 \left(\frac{x^2 y}{4} \right)^{2/3}.$$

Putting things together, we have

$$4 = x + y + \sqrt{2x^2 + 2xy + 3y^2}$$

$$\geq 3 \left(\frac{x^2 y}{4} \right)^{1/3} + \sqrt{15} \left(\frac{x^2 y}{4} \right)^{1/3}$$

$$> 4 \left(\frac{x^2 y}{4} \right)^{1/3},$$

(unless $x = 0$ or $y = 0$, in which case the claim is obviously valid), giving

$$1 > \left(\frac{x^2 y}{4} \right)^{1/3},$$

and

$$4 > x^2 y.$$

GLEANINGS FROM MURRAY KLAMKIN'S OLYMPIAD CORNERS—1982

1. West German Olympiad, 1982, First Round (p. 70)

1. (revised slightly).   What is the *sum* of the *greatest odd* divisors of the integers $1, 2, 3, \ldots, 2^n$?

In general, let $S(a, b, \ldots)$ denote the sum of the greatest odd divisors of $a, b, \ldots$, and, in particular, let $S_n = S(1, 2, 3, \ldots, 2^n)$; we desire a formula for S_n.

Clearly, the greatest odd divisor of an odd integer m is m itself, and the greatest odd divisor of $2k$ is the same as that of k. Accordingly, we have

$$S_n = S(1, 2, 3, \ldots, 2^n)$$

$$= S(1, 3, 5, \ldots, 2^n - 1) + S(2, 4, 6, \ldots, 2^n)$$

$$= (1 + 3 + 5 + \cdots + 2^n - 1) + S(1, 2, 3, \ldots, 2^{n-1}).$$

But the sum of the first k positive odd integers is well known to be just k^2. Since $2^n - 1$ is the 2^{n-1}th odd positive integer, then

$$S_n = (2^{n-1})^2 + S(1, 2, 3, \ldots, 2^{n-1})$$

$$= 4^{n-1} + S_{n-1},$$

giving

$$S_n - S_{n-1} = 4^{n-1}.$$

Repeatedly using this result, we get

$$S_n - S_{n-1} = 4^{n-1}$$
$$S_{n-1} - S_{n-2} = 4^{n-2}$$
$$S_{n-2} - S_{n-3} = 4^{n-3}$$
$$\vdots$$
$$S_2 - S_1 = 4$$

(one must always double check to be sure the final line is within the range of validity of the relations; since $S_2 = 6$ and $S_1 = 2$, we are safe). Adding gives

$$S_n - S_1 = 4 + 4^2 + \cdots + 4^{n-1}$$
$$= \frac{4(4^{n-1} - 1)}{4 - 1},$$

and

$$S_n = 2 + \frac{4^n - 4}{3} = \frac{4^n + 2}{3}.$$

2. The All-Russian Olympiad, 1979–80, Grade 9 (p. 72)

4.   In converting the fraction m/n to a decimal by long division, where m and n are positive integers and n does not exceed 100, young Vladimir came up with a quotient which contained, somewhere after the decimal point, the consecutive digits 167, in that order. Prove that Vladimir must have made a mistake.

$$q = \ldots \cdot \overbrace{\ldots 167}^{k} \ldots$$
$$n\sqrt{\ldots m \ldots \cdot 00 \ldots 00000 \ldots}$$

$$\ddots$$

$$\ldots$$

$$\underline{(7n)}$$

$$r0$$

$$\ddots$$

When the long division has been completed to the point of getting the 167 in the quotient q, let the remainder be r, and suppose the number of decimal places in q at this time is k. At this point the remainder represents units of size $1/10^k$ and the long division to this stage amounts to the equation

$$m = qn + \frac{r}{10^k}, \quad \text{or} \quad 10^k m = 10^k qn + r;$$

thus, for some integer $t \geq 0$, we have

$$10^k m = 10^k (\ldots \cdot \overbrace{\ldots 167}^{k \text{ places}}) n + r$$
$$= \ldots 167n + r$$
$$= (1000t + 167)n + r$$
$$= 1000tn + 167n + r.$$

Since q contains at least the digits 167 after the decimal point, k must be at least 3, and therefore $1000 | 10^k$. Hence 1000 must also divide $167n + r$, and, for some integer s, we have $167n + r = 1000s$. Since $n \leq 100$, and the remainder r is even less than n, we have $0 \leq r < 100$, and

$$167n + r < 16700 + 100 = 16800,$$

implying s is at most 16, that is, $s \in \{1, 2, 3, \ldots, 16\}$.

Since $0 \leq r < 100$, the number $167n = 1000s - r$ must occur somewhere in the range $(1000s - 100, 1000s]$. But, because $6(167) =$

1002 is *so close to* 1000, this requirement is *not* met by any value of $n \leq 100$: with $s \in \{1, 2, 3, \ldots, 16\}$, the possible ranges are

$$(900, 1000], (1900, 2000], (2900, 3000], \ldots, (15900, 16000];$$

with $6(167) = 1002$, the multiples of 167 jump right over all these ranges:

$$6(167) = 1002, \quad \text{and the previous multiple is } 835$$

$$12(167) = 2004, \quad \text{and the previous multiple is } 1837,$$

$$18(167) = 3006, \quad \text{and the previous multiple is } 2839,$$

$$\cdots$$

$$96(167) = 16032, \quad \text{and the previous multiple is } 15865.$$

Thus young Vladimir must have made a mistake somewhere!

3. Bulgarian Olympiad, 1982 (p. 237)

In the plane, n circles of unit radius are drawn with different centers. Of course, overlapping circles partly cover each other's circumferences. A given circle could be so overlaid that any uncovered parts of its circumference are all quite small; that is, it might have no sizable uncovered arcs at all. However, this can't be true of every circle; prove that some circle must have a continuously uncovered arc which is at least $1/n$th of its circumference.

The convex hull H of the given set of circles will be a closed convex figure whose boundary B consists of segments and of arcs of the given unit circles. In going all the way around B, one turns altogether through a complete revolution. No turning takes place along the "straightaways," only around the arcs; thus, besides the straight sections, one also traverses exactly one complete circumference of a unit circle.

Since there are only n circles, B cannot contain more than n arcs, implying that the *average* length of an arc is at least $1/n$th a circumference. Since all arcs can't be below average, some arc of B must be at least $1/n$th a circumference and, being so located, is obviously not intersected by any other circle.

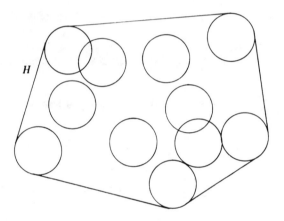

FIGURE 58

4. Peking Math Contest, 1962, Grade 12, Paper 2 (p. 269)

What is the value of the sum

$$S = m! + \frac{(m+1)!}{1!} + \frac{(m+2)!}{2!} + \cdots + \frac{(m+n)!}{n!} \quad ?$$

We have

$$S = m! \left[1 + \frac{(m+1)!}{m!1!} + \frac{(m+2)!}{m!2!} + \cdots + \frac{(m+n)!}{m!n!} \right]$$

$$= m! \left[\binom{m}{m} + \binom{m+1}{m} + \binom{m+2}{m} + \cdots + \binom{m+n}{n} \right].$$

Although the sum in the bracket is an awkward expression from an algebraic standpoint, the following combinatorial interpretation is a very neat way of showing that its value is just $\binom{m+n+1}{m+1}$.

Suppose you want to count the number of ways of selecting a subset A of $m+1$ integers from the set $\{1,2,3,\ldots,m+n+1\}$; clearly, there is a total of $\binom{m+n+1}{m+1}$ ways of doing it.

Now the greatest integer k that is selected in a subset A can not be less than $m+1$, the number of integers in A, nor bigger than $m+n+1$, the maximum integer available; and it can be any integer in between.

(i) If $k = m+1$, then the m smaller members of A must come from the segment $\{1,2,3,\ldots,m\}$, and there are $\binom{m}{m}$ ways of constructing such an A;

(ii) if $k = m+2$, the m smaller members come from $\{1,2,\ldots,m+1\}$, and they can be chosen in $\binom{m+1}{m}$ ways;

. . . .

(iii) if $k = m+n+1$, the others come from $\{1,2,\ldots,m+n\}$, and can be chosen in $\binom{m+n}{m}$ ways,

for a grand total of

$$\left[\binom{m}{m} + \binom{m+1}{m} + \cdots + \binom{m+n}{m}\right]$$

ways.

Accordingly,

$$S = m!\binom{m+n+1}{m+1} = m!\frac{(m+n+1)!}{(m+1)!n!},$$

or finally,

$$S = \frac{(m+n+1)!}{(m+1)\cdot n!}.$$

5. Peking Math Contest, 1963, Grade 12 (p. 269)

4.   To begin, a set of 2^n objects is partitioned into an arbitrary number of nonempty subsets. Then the subsets are permitted

to be combined two at a time as follows: first their cardinalities are compared—

(i) if they are equal, the subsets are simply pooled into one subset of twice the size,

(ii) otherwise a transfer is made from the subset of larger cardinality to the subset of lesser cardinality so as to double the size of the recipient.

Prove that, proceeding judiciously, it is possible to combine all the subsets into a single set in a finite number of moves.

For example, let $2^5 = 32$ objects be partitioned into subsets of sizes 10, 9, 8, and 5. One might proceed as follows:

(i)   combine 9 and 5 to give 4 and 10 and the subsets 10, 4, 8, 10;

(ii) combine 10 and 10 to obtain subsets 20, 4, 8;

(iii) combine 20 and 4 to give 16, 8, 8;

(iv) combine 8 and 8 to give 16, 16;

(v)   combine 16 and 16 to complete the reassembly.

Solution by Willie Yong, Singapore.   As you might expect, the proof is by induction. Clearly, the only partition of $2^1 = 2$ objects is into two singleton subsets, which restore the set in one move. Suppose, then, that any set of 2^{n-1} objects can be restored from any partition in a finite number of moves, and let us consider a given partition of a set of 2^n objects.

First of all Willie observes that the *number* of subsets which contain an *odd* number of elements must be an *even* number (otherwise the total number of objects couldn't be even and equal to 2^n). Setting aside the other sets (each of which has an *even* number of elements), he concentrates on combining pairs of these "odd" subsets. Because a move doubles the size of one of the participating subsets, the combination of two odd subsets must result in two even subsets (one of which could be empty if equal subsets are combined; but the empty set is trivially even). In any event, the combination of two odd subsets reduces the number of odd subsets by 2. Since there is an even number of them to begin with, they can be paired off and completely eliminated 2 at a time, to leave a collection of nothing but even subsets.

Now, the elements of an even subset can be arbitrarily paired off among themselves to yield a subset whose elements are the resulting pairs and whose cardinality is just half of what it was (e.g., a subset of 40 objects gives a new subset of 20 pairs). Doing this with all the current subsets, which are even, provides a collection of subsets of these pairs which is really a partition of the set of 2^{n-1} pairs that was just formed from the original 2^n objects. By proceeding with these subsets as prescribed by the induction hypothesis, one arrives at a single subset containing all the pairs and which therefore also contains all 2^n of the original objects.

6. West German Olympiad, 1982 (p. 300)

4. If a positive integer n makes $4^n + 2^n + 1$ a prime number, prove that n must be a power of 3.

Solution by Ken Davidson, University of Waterloo. Let us prove the equivalent contrapositive statement: if n is *not* a power of 3, then $4^n + 2^n + 1$ is **not** a prime.

Suppose that n is not a power of 3. This means that n has a prime divisor $p \neq 3$ and that, for some complementary divisor k, we have $n = kp$, k a positive integer. Then

$$4^n + 2^n + 1 = 4^{kp} + 2^{kp} + 1 = \left(2^{kp}\right)^2 + 2^{kp} + 1$$
$$= \frac{\left(2^{kp}\right)^3 - 1}{2^{kp} - 1},$$

and we have $(2^{kp} - 1)[(2^{kp})^2 + 2^{kp} + 1] = (2^{kp})^3 - 1$. Now the right side can be factored alternatively in the form

$$\left(2^{kp}\right)^3 - 1 = \left(2^{3k}\right)^p - 1^p = \left(2^{3k} - 1\right)(\cdots),$$

where the second factor need not be specified. Thus we have that

$$2^{3k} - 1 \bigg| \left(2^{kp} - 1\right)\left[\left(2^{kp}\right)^2 + 2^{kp} + 1\right].$$

If we can show that $2^{3k} - 1$ fails to divide completely into $2^{kp} - 1$, it must be that some nontrivial factor of $2^{3k} - 1$ must divide into the other part, namely $(2^{kp})^2 + 2^{kp} + 1$. Because p is a prime, $2p > 3$, implying that

$$(2^{kp})^2 + 2^{kp} + 1 > 2^{3k} - 1.$$

Therefore, any factor $f > 1$ of $2^{3k} - 1$ that divides $(2^{kp})^2 + 2^{kp} + 1$ would be a nontrivial *proper* divisor, showing that $(2^{kp})^2 + 2^{kp} + 1$ is *not* a prime, as desired.

To show that $2^{3k} - 1$ fails to divide $2^{kp} - 1$, we observe, because the prime p is not divisible by 3, that $p = 3t + r$, where $r = 1$ or 2, t some positive integer. Then

$$2^{kp} - 1 = 2^{k(3t+r)} - 1 = 2^{kr} \cdot 2^{3kt} - 1.$$

Peeling off one of these 2^{3kt}'s to go with the -1, we have

$$2^{kp} - 1 = \left(2^{kr} - 1\right) \cdot 2^{3kt} + \left(2^{3kt} - 1\right).$$

Now, because $2^{3kt} - 1 = (2^{3k})^t - 1^t = (2^{3k} - 1)(\cdots)$, it follows that $2^{3k} - 1 | 2^{3kt} - 1$. Therefore $2^{3k} - 1$ can only divide $2^{kp} - 1$ if it also divides $(2^{kr} - 1) \cdot 2^{3kt}$; and since $2^{3k} - 1$ is odd, this means it must also divide $2^{kr} - 1$. But this is impossible since $r = 1$ or 2, making $2^{kr} - 1$ *less* than the proposed divisor $2^{3k} - 1$. Thus $2^{3k} - 1$ *does* fail to divide $2^{kp} - 1$, and our argument is complete.

A QUADRUPLE OF CONSECUTIVE INTEGERS

The numbers $(160, 161, 162)$ constitute a set of three consecutive integers that are divisible, respectively, by 5, 7, and 9. Determine a set of four consecutive positive integers that are divisible, respectively, by 5, 7, 9, and 11.

Solution

Since the last number is divisible by 11, the required set has the form

$$\{11k - 3, 11k - 2, 11k - 1, 11k\},$$

that is

$$\{11(k + 2) - 25, 11(k + 3) - 35, 11(k + 4) - 45, 11k\},$$

and we see that we would like to find k such that

$$5|k + 2, \quad 7|k + 3, \quad \text{and} \quad 9|k + 4.$$

Since $k + 2$, $k + 3$, and $k + 4$ are consecutive, we can make use of the example given in the question and take

$$k + 2 = 160, \quad \text{giving} \quad k = 158, \quad 11k = 1738,$$

and the quadruple

$$\{1735, 1736, 1737, 1738\}.$$

Of course, a genius would have seen at a glance that, for

$$k = 5 \cdot 7 \cdot 9 \cdot 11 = 3465,$$

the quantities

$$\frac{k+5}{2}, \quad \frac{k+7}{2}, \quad \frac{k+9}{2}, \quad \frac{k+11}{2}$$

are all *integers* since k is odd, *consecutive* because they increase from one to the next by 2/2, and that dividing by the denominator 2 does not affect any *odd* divisor of the numerator.

A BOX-PACKING PROBLEM

A 3-brick is a $1 \times 1 \times 3$ block of 3 unit cubes in a row. In a $7 \times 7 \times 7$ box B there is room for $7^3 = 343$ unit cubes. Since 343 is not divisible by 3, B cannot be packed exactly with just 3-bricks. However, 114 3-bricks can be packed into B so as to occupy all the unit cells except one; that is, there will be a single unit-cell hole H someplace. In fact B can be packed like this in lots of ways; the illustration shows the hole H at a corner. H could also go in the middle of an edge or at the center of a face. However, it is curious that if H is not on the outside surface, but buried somewhere in the $5 \times 5 \times 5$ inner core, then the only place it can possibly be is the very center cell itself; prove this.

Solution

The basis of our approach is a scheme for coloring the 343 unit cells in the box B. Each cell is assigned one of three colors, 1, 2, 3, in such a way that, no matter where a 3-brick is packed into the box, its 3 cells will occupy one cell of each color.

A reasonably regular coloring with this property can be obtained from copies of a $3 \times 3 \times 3$ block A that is colored as shown in Figure 60.

133

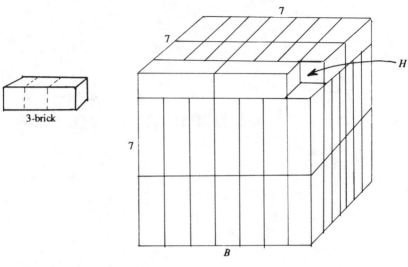

FIGURE 59

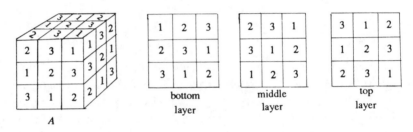

FIGURE 60

Let 27 copies of A be stacked in a $3 \times 3 \times 3$ array, each copy oriented toward us as pictured in the figure (that is, without any turning), to form a $9 \times 9 \times 9$ cube. Then if the top two layers, the front two slices, and the two slices on the right side are sheared off, a properly colored $7 \times 7 \times 7$ box B remains, colored as illustrated in Figure 61.

Clearly this makes layers 1, 4, and 7 the same, layers 2 and 5 the same, and layers 3 and 6 the same. It is not difficult to check that wherever a 3-brick is placed, it occupies a cell of each color. Thus, altogether the 114 3-bricks will occupy 114 cells of each color. Counting up the cells of each color, we find that there are *only* 114 each of colors 2 and 3, and 115 of color 1. The hole H, therefore, must occur at a cell colored 1.

1	2	3	1	2	3	1
2	3	1^P	2	3	1	2
3	1	2	3	1	2^S	3
1	2	3	1	2	3	1
2	3^Q	1	2	3	1	2
3	1	2	3	1^R	2	3
1	2	3	1	2	3	1

layers 1, 4, 7

2	3	1	2	3	1	2
3	1	2	3	1	2	3
1	2	3	1	2	3	1
2	3	1	2	3	1	2
3	1	2	3	1	2	3
1	2	3	1	2	3	1
2	3	1	2	3	1	2

layers 2, 5

3	1	2	3	1	2	3
1	2	3	1	2	3	1
2	3	1	2	3	1	2
3	1	2	3	1	2	3
1	2	3	1	2	3	1
2	3	1	2	3	1	2
3	1	2	3	1	2	3

layers 3, 6

B

FIGURE 61

A quick glance shows that this still leaves lots of places for H in the inner core. However, if B can be packed with H in some cell C, then, by simply spinning B through turns of one-, two-, and three-quarters of a revolution about its three axes of symmetry, we see that it can also be packed with H in any of the cells into which C is carried by these rotations. For example, if B could be packed with H in the cell marked P of layer 4 (see Figure 61), then it could also be packed with H at any of Q, R, and S (and also two other sets of symmetric cells in perpendicular slices). Now, unless *all* of these symmetric cells have color 1, H cannot occur at any of them. For example, H cannot go at Q, since Q is not color 1; but *if* H could go at P, then it could also go at Q, giving a contradiction. Hence H cannot go at P either, even though it is colored 1.

Thus, if H is to occur in the $5 \times 5 \times 5$ inner core, it must be at the very center, for it is easily checked that this is the only cell colored 1 in the core all of whose symmetric cells are also colored 1.

AN AWKWARD INTEGRAL

Evaluate the integral

$$I = \int_0^{\pi/2} \frac{\sin^{25} x}{\cos^{25} x + \sin^{25} x}\, dx.$$

Solution

The values taken by this awkward trigonometric integrand $f(x)$ as x runs from 0 to $\pi/2$ constitute the same set of values as those taken by $f(\pi/2 - x)$ as x goes from 0 to $\pi/2$, except that they are generated in reverse order. Consequently, the integral is unchanged by replacing $f(x)$ by $f(\pi/2 - x)$, and we have

$$I = J = \int_0^{\pi/2} \frac{\sin^{25}\left(\frac{\pi}{2} - x\right)}{\cos^{25}\left(\frac{\pi}{2} - x\right) + \sin^{25}\left(\frac{\pi}{2} - x\right)}\, dx$$

$$= \int_0^{\pi/2} \frac{\cos^{25} x}{\sin^{25} x + \cos^{25} x}\, dx.$$

Now who could resist putting I and J together:

$$I + J = \int_0^{\pi/2} 1 \cdot dx = \frac{\pi}{2}.$$

Since $I = J$, then

$$I = \frac{\pi}{4}.$$

A MATTER OF PERSPECTIVE

AB and CD are consecutive ties across railway tracks that appear to meet at O on the horizon H. If the ties are parallel to H and are equally spaced along the tracks, how do you draw the next tie in this perspective figure?

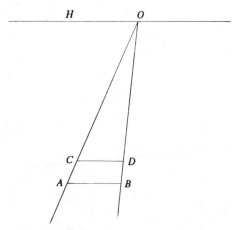

FIGURE 62

Solution

Two straight lines which meet on the horizon H (that is, the line at infinity) are parallel. The regions between the tracks and consecutive ties are identical rectangles. Thus, if the required tie is EF, the diagonals BC and DE are parallel. Now if BC meets H at P, then PD is the line through D which is parallel to BC. Accordingly, PD crosses the first rail at the required point E, and it remains only to draw EF parallel to CD.

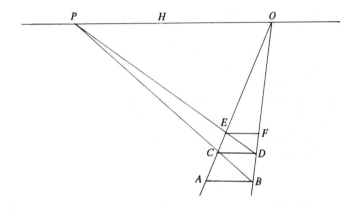

FIGURE 63

SEQUENCES OF NESTED RADICALS

Now let's consider a problem about some interesting infinite sequences of positive real numbers which are given in terms of nested radicals. Suppose the sequences $\{a_n\}, \{b_n\}$, and $\{c_n\}$ are defined for the positive integers as follows:

(a)

$$a_1 = \sqrt{1}, \quad a_2 = \sqrt{1 + \sqrt{1}},$$

$$a_3 = \sqrt{1 + \sqrt{1 + \sqrt{1}}}, \ldots,$$

$$a_n = \sqrt{1 + \sqrt{1 + \sqrt{1 + \ldots + \sqrt{1}}}}, \quad \text{with } n \text{ 1's;}$$

(b)

$$b_1 = \sqrt{1}, \quad b_2 = \sqrt{1 + \sqrt{2}},$$

$$b_3 = \sqrt{1 + \sqrt{2 + \sqrt{3}}}, \ldots,$$

$$b_n = \sqrt{1 + \sqrt{2 + \ldots + \sqrt{n}}};$$

(c)

$$c_1 = \sqrt{1}, \quad c_2 = \sqrt{1 + \sqrt{2 + \sqrt{2^2}}},$$

$$c_3 = \sqrt{1 + \sqrt{3 + \sqrt{3^2 + \sqrt{3^3}}}}, \ldots,$$

$$c_n = \sqrt{1 + \sqrt{n + \sqrt{n^2 + \ldots + \sqrt{n^n}}}}.$$

The problem is:

(i) to prove that $\{a_n\}$ converges and to determine its limit L;
(ii) to determine whether $\{b_n\}$ is convergent; and
(iii) to show that $c_n < n$ for all $n > 1$.

Solution

(i)   Since a_{n+1} is obtained from a_n by increasing its final 1 to $1 + \sqrt{1}$, we have at once that $\{a_n\}$ is strictly monotonically increasing. The matter of its convergence is settled by showing that it is bounded. A standard approach to this problem is to calculate the first few terms of the sequence, try to guess a reasonable bound, and then try to prove that your guess really *is* a bound. Although guessing lies at the heart and soul of mathematics, you might prefer not to resort to it when you don't have to. The wording of the question guarantees that $\{a_n\}$ is convergent (we don't deal in deliberately misleading questions), and because it is monotonic, its limit L is automatically a bound on it. Therefore the crafty thing to do is to proceed with the computation of L, as if we knew already that the sequence was convergent, and then try to establish a bound on a_n that we can choose for its convenience and still be sure of its feasibility. At the end of this, if all goes well, a slight rearrangement of our work will yield the two steps necessary for the establishment of the convergence of $\{a_n\}$, and we shall also have in hand the required limit L as a bonus.

With the obvious relation $a_n = \sqrt{1 + a_{n-1}}$, we proceed accordingly with

$$L = \lim_{n \to \infty} a_n = \lim_{n \to \infty} \sqrt{1 + a_{n-1}} = \sqrt{1 + \lim_{n \to \infty} a_{n-1}} = \sqrt{1 + L},$$

giving $L^2 = 1 + L$, $L^2 - L - 1 = 0$, and $L = (1 \pm \sqrt{5})/2$. Since L is positive, then L is none other than the famous

Golden Ratio: $L = g = \dfrac{1 + \sqrt{5}}{2} = 1.618\ldots$.

Consequently, a_n is bounded by every number $\geq 1.618\ldots$. Therefore we can be sure that a_n is always < 2. Let's see if we can find an easy proof of this (if not, we can slacken off to other nice values like $3, 4, \ldots$, and try again). From $a_n = \sqrt{1 + a_{n-1}}$, it is clear that

if $a_{n-1} < 2$,   then $a_n < \sqrt{1 + 2} = \sqrt{3} < 2$.

Since $a_1 = 1 < 2$, then $\{a_n\}$ is indeed bounded, by induction, and our solution for part (i) is complete. We note that $a_n < g$ for all n.

(ii)   Since b_{n+1} is obtained from b_n by replacing n by $(n + \sqrt{n+1})$, we again have that the sequence is strictly monotonically increasing. Again, the issue is whether $\{b_n\}$ is bounded.

The first few terms strongly suggest that it *is*, even that it is again always less than 2:

$$b_1 = 1, \qquad\qquad b_2 = 1.5538\ldots,$$

$$b_3 = 1.7123\ldots, \qquad b_4 = 1.7488\ldots,$$

$$b_5 = 1.7562\ldots, \qquad b_6 = 1.7576\ldots,$$

$$b_7 = 1.7579\ldots, \qquad b_8 = 1.7579\ldots.$$

Unfortunately, there is no simple relation connecting b_n and b_{n-1}, like the $a_n = \sqrt{1 + a_{n-1}}$ in part (i), and so proving $b_n < 2$ is not the simple task we enjoyed above. What we need this time is some way of showing that the potentially large contribution made by the number n to the value of

$$b_n = \sqrt{1 + \sqrt{2 + \sqrt{3 + \cdots + \sqrt{n}}}}$$

is nullified by the many times the square root must be taken. This is brilliantly achieved at one stroke by the unlikely maneuver of removing a factor of $\sqrt{2}$:

$$b_n = \sqrt{2} \cdot \sqrt{\frac{1}{2} + \sqrt{\frac{2}{2^2} + \sqrt{\frac{3}{2^4} + \cdots + \sqrt{\frac{n}{2^{2^{n-1}}}}}}}.$$

Clearly, then, we have

$$b_n < \sqrt{2} \cdot \sqrt{1 + \sqrt{1 + \cdots + \sqrt{1}}}$$
$$= \sqrt{2} \cdot a_n \qquad \text{(the } a_n \text{ from part (i))}$$
$$< \sqrt{2}g,$$

and the convergence of $\{b_n\}$ is established.

(iii) Although $\{c_n\}$ is a more complicated sequence, the same bold approach makes it anticlimactic. Treating

$$c_n = \sqrt{1 + \sqrt{n + \sqrt{n^2 + \cdots + \sqrt{n^n}}}}$$

in the same way, we obtain

$$c_n = \sqrt{n} \cdot \sqrt{\frac{1}{n} + \sqrt{\frac{n}{n^2} + \sqrt{\frac{n^2}{n^4} + \cdots + \sqrt{\frac{n^n}{n^{2^n}}}}}}$$
$$< \sqrt{n}\sqrt{1 + \sqrt{1 + \cdots + \sqrt{1}}} < \sqrt{n}g.$$

Hence the required $c_n < n$ holds for

$$\sqrt{n}g < n,$$

$$g^2 < n,$$

that is,

$$n > 2.618\ldots,$$

$$n \geq 3.$$

Since

$$c_2 = \sqrt{1 + \sqrt{2 + \sqrt{2^2}}} = \sqrt{3} < 2,$$

then

$$c_n < n \qquad \text{for all } n > 1.$$

EQUATIONS OF FACTORIALS

The factorials attain astronomical proportions so quickly that our minds are soon overwhelmed by them. Thus it is with a certain reticence that we might approach an equation which contains only factorials.

> For positive integers x, y, z, prove that there is *only one* solution to $x! + y! = z!$, but that the equation $x!y! = z!$ has *infinitely many* solutions with $x, y, z, > 1$ $((x, y, z) = (1, y, y)$ is a trivial solution for all y).

Solution

(i) Since $1! = 1$ and $2! = 2$, we have immediately that $(x, y, z) = (1, 1, 2)$ is a solution of $x! + y! = z!$. Since x and y are both positive, we must have $z > x, y$. Thus for $z = 2$, there is no choice but to have $x = y = 1$, making $(1, 1, 2)$ the *only* solution with $z = 2$.

Also from $x, y < z$, we have x and y no bigger than $z - 1$; thus, for $z > 2$, we have

$$x! + y! \leq 2(z - 1)! < z(z - 1)! = z!.$$

Hence there is no solution with $z > 2$, making $(1, 1, 2)$ the *unique* solution of $x! + y! = z!$.

The equation $x! \cdot y! = z!$ presents us with the opposite kind
m—that of providing an endless supply of solutions.

ght take some time for the significance of the fact that both
iis equation are *products* to make itself felt. Thus the problem
can be viewed as that of starting with $z!$ and factoring it into two factorials. The often encountered "peeling off the top factor" decomposition
of a factorial is bound to cross one's mind sooner or later and we should
then be faced with

$$z \cdot (z - 1)! = z!.$$

(In fact, we encountered this above just moments ago.) From here it
remains only to choose z itself from among the factorials. Any factorial
will do, and it will do double duty by also serving for the value of $x!$,
yielding the solution $(x, x! - 1, x!)$:

$$x! \cdot y! = x! \cdot (x! - 1)! = (x!)! = z!.$$

For example, $x = 6$, gives the solution $(6, 6! - 1, 6!) = (6, 719, 720)$:

$$6! \cdot (6! - 1)! = (6!)!,$$

i.e.,

$$720 \cdot 719! = 720!.$$

GLEANINGS FROM MURRAY KLAMKIN'S
OLYMPIAD CORNERS—1983

1. Part (a), Problem K-2, from Kvant (p 72)

The first n positive integers $(1, 2, 3, \ldots, n)$ are spotted around a circle in any order you wish and the positive differences $d_1, d_2, \ldots, d_n$ between consecutive pairs are determined. Prove that, no matter how the integers might be jumbled up around the circle, the sum of these n differences,

$$S = d_1 + d_2 + \cdots + d_n,$$

will always amount to at least $2n - 2$.

The following incisive solution is due to Noam Elkies, at the time a student at Columbia University.

If the integers around the circle are $a_1, a_2, \ldots, a_n$, the problem involves the awkward sum of absolute values

$$S = |a_1 - a_2| + |a_2 - a_3| + |a_3 - a_4| + \cdots + |a_n - a_1|.$$

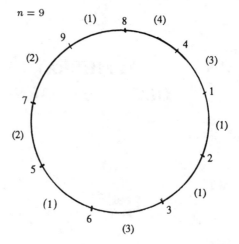

FIGURE 64

$S = 4 + 3 + 1 + 1 + 3 + 1 + 2 + 2 + 1 = 18 \geq 2(9) - 2 = 16.$

Fortunately, we are only required to establish a given lower bound on this sum.

Now, the integers 1 and n have to occur somewhere on the circle. Consider the sum S_1 of the differences that are generated by the integers that lie along one of the arcs between the n and the 1:

$$S_1 = |n - x| + |x - y| + |y - z| + \cdots + |p - q| + |q - 1|.$$

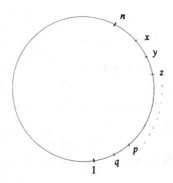

FIGURE 65

Clearly we have in general that

$$|a + b + c + \cdots| \leq |a| + |b| + |c| + \cdots,$$

and so

$$S_1 \geq |n - x + x - y + y - z + \cdots + p - q + q - 1|$$
$$= |n - 1| = n - 1.$$

Since the complementary arc yields a similar sum $S_2 \geq n - 1$, we have

$$S = S_1 + S_2 \geq 2(n - 1) = 2n - 2.$$

2. Bulgarian Winter Competition, 1983

Prove that a regular hexagon $H = ABCDEF$ of side 2 can be covered completely with 6 disks of unit radius, *but not by 5.*

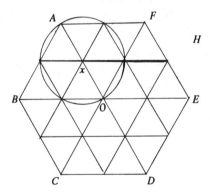

FIGURE 66

Clearly the segments $OA, OB, OC, \ldots$, from the center O to the vertices, partition H into 6 equilateral triangles of side 2, and the segments joining the midpoints of their sides further divide H into 24 equilateral triangles of side 1. It is evident that a unit disk with center at the midpoint X of OA nicely covers the 6 such triangles around X, and that

5 similar disks, with centers at the midpoints of $OB, OC, \ldots$, provide complete coverage of H.

In fact, in this arrangement 12 of the little triangles get covered twice, showing that 6 disks provide a great deal more coverage than the area of H requires. Our main problem is to show that, although 5 disks still have plenty of raw covering power, their own shapes and that of H more than exhaust their collective covering capabilities.

The boundary of H has a total length of 12 units, and if it were possible to cover H with just 5 unit disks, they would have to cover an average of 12/5 units of boundary apiece. If we could show that no unit disk is capable of covering this much of the boundary, our desired conclusion would follow.

If a disk covers only one side of H, then it could not cover a section longer than its 2-unit diameter. Clearly a disk is not big enough to intersect simultaneously 3 of the sides of H. Consider, then, a disk that covers parts of two adjacent sides of H. If the sides meet at the vertex V, then there are two cases, depending on whether the disk also covers V.

(i)   Suppose that V is also covered by the disk (see Figure 67).

First we observe that the angle at each vertex of H is $60 + 60 = 120$ degrees. Now, let K be the circle through the vertex V and the endpoints X and Y of the intercepts on the sides in question, and let Z be the midpoint of the arc XY along K; let R be the circle with center Z and radius $ZX = ZY$, and let XV be extended to meet R at W.

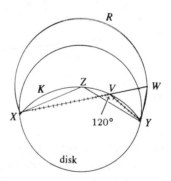

FIGURE 67

Then, in K, the chord XY subtends the same angle at V and Z, making $\angle XZY = 120°$. But Z is the center of the circle R, and therefore the angle XY subtends at W on the circumference is half as great, making $\angle XWY = 60°$. Since exterior angle XVY of $\triangle VWY$ is equal to the sum of the interior angles at Y and W, we have that

$$\angle VYW = 120 - 60 = 60°.$$

This makes $VY = VW$, and the total intercept in question is

$$XV + VY = XV + VW = XW,$$

a *chord in the circle R*. But no chord in R can exceed its diameter; hence

$$XV + VY = XW \leq \text{ diameter of } R = 2 \cdot XZ.$$

That is to say, the total intercept cannot exceed twice ZX, where XYZ is an isosceles triangle having a vertical angle of $120°$ at Z. By the law of cosines, we have

$$XY^2 = XZ^2 + YZ^2 - 2XZ \cdot YZ \cos 120°$$
$$= 2XZ^2 - 2XZ^2(-\frac{1}{2}) = 3XZ^2.$$

Because XY is a chord of our unit disk, we have

$$XY \leq \text{ the diameter 2,}$$

and

$$3XZ^2 = XY^2 \leq 4,$$

giving

$$XZ \leq \frac{2}{\sqrt{3}},$$

and

$$2XZ \le \frac{4}{\sqrt{3}} = \frac{4\sqrt{3}}{3}$$

$$< \frac{4(1 \cdot 74)}{3}$$

$$< 2 \cdot 33 < 2 \cdot 4 = \frac{12}{5}.$$

(ii)   Suppose, finally, that V is not covered by the disk.

Let the intercepts on the sides be PX and QY, and let PZ be drawn parallel to VY to meet the disk at Z, as shown. Then $\angle XPZ = \angle XVY = 120°$, and we have

$$XP + QY < XP + VY$$

$$< XP + PZ$$

$$< \frac{12}{5}, \quad \text{by part (i).}$$

Hence no disk can intercept as much as 12/5 units of the boundary of H, implying that 5 disks are not sufficient to cover the entire extent of H.

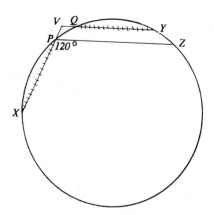

FIGURE 68

3. British Olympiad, 1983 (p. 108)

If 10 points are chosen in a circle C of diameter 5, prove that the distance between some pair of them is less than 2.

Having encountered problems like this in Morsel 17 and in Section 9 of the 1981 Gleanings, one's first thought might be to see whether the same approach will work again. Accordingly, let a circle of unit radius be drawn with each of the 10 chosen points as center; clearly, if two of these circles overlap in interior points, then the distance between their centers is less than 2.

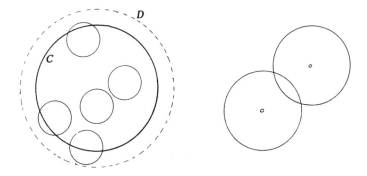

FIGURE **69**

Although the unit circles might extend beyond the edge of C, they are all contained by a concentric circle D of diameter 7. Now, two of our unit circles would certainly intersect in a region having positive area, i.e., in interior points, if the total area of the 10 circles were to exceed the area of D. Checking, we find that

$$\text{the area of } D = \pi(\frac{7}{2})^2 = \frac{49}{4}\pi > 12\pi,$$

but the area of 10 unit circles is only 10π. It was a nice thought, but there is nothing to do now but accept this little setback with good humor and take another run at the problem.

Even though the above application of the pigeonhole principle didn't work out, surely the pigeonhole principle must be the basis of any solution.

No matter how the given circle might be partitioned into **9** regions, the pigeonhole principle implies that some 2 of the 10 chosen points have to belong to the same region. Consequently, if we were able to show that there exists a partition of C into 9 regions, *each of diameter less than* 2, the desired conclusion would follow.

There is no reason why all 9 regions have to be the same shape; we are only concerned that the diameter be less than 2. In this case, let's begin with a unit circle K, concentric with C, and a set of 8 spokes, radiating from the center 0 at angles of 45°, so that the annulus between K and C is divided into 8 congruent regions R; also, let coordinate axes be assigned, as shown. (Recall section 3 of Gleanings, 1981.)

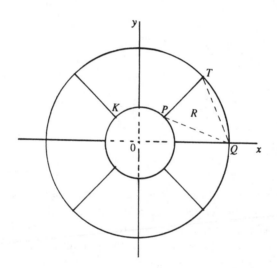

FIGURE 70

Clearly, the diameter of a region R is given either by the chord QT or by the diagonal PQ. Showing that both QT and PQ are less than 2 will establish that the diameter of R is less than 2. We have

$$QT < \text{arc } QT = \frac{\text{circumference of } C}{8} = \frac{5\pi}{8} = \frac{10\pi}{16} < 2, \quad \text{(i)}$$

and, noting that P is the point $(\cos 45°, \sin 45°)$ and Q is $(5/2, 0)$,

$$PQ = \sqrt{(\frac{5}{2} - \cos 45°)^2 + \sin^2 45°}$$

$$= \sqrt{\frac{25}{4} - 5\cos 45° + 1} \quad \text{(ii)}$$

$$= \sqrt{\frac{29}{4} - \frac{5}{\sqrt{2}}} = \sqrt{\frac{29 - 10\sqrt{2}}{4}} < \frac{1}{2}\sqrt{29 - 14}$$

i.e., $PQ < \frac{1}{2}\sqrt{15} < \frac{1}{2}\sqrt{16} = 2$. Thus the diameter of a region R is less than 2 alright.

There is one little hitch, however. While each R has diameter less than 2, the unit circle K has diameter *exactly* 2. Therefore, to obtain our final partition of C, let K be shrunk, about the same center O, to a radius of $1 - \epsilon$, slightly less than 1. Of course, this extends the boundary of all the regions R, increasing their diameters. But, no matter how little the diameter of R might have fallen short of 2, ϵ can be chosen to be so small that the increased diameter of the new R will still be < 2. However, any small positive ϵ yields a new circle K of diameter $2(1 - \epsilon) < 2$. The existence of a suitable ϵ shows that there *does* exist a partition of C into 9 regions, each of diameter strictly less than 2, and our argument is complete.

4. American Invitational Mathematics Examination, 1983 (p. 171)

13.   Let A be any subset of $N = \{1, 2, 3, \ldots, n\}$, and let the members of A be arranged in decreasing order of magnitude. Now form a sum S by alternately adding and subtracting successive members in this arrangement. For example, for $n = 20$, the subset $A = \{11, 6, 17, 1, 9, 18, 13\}$ yields the sum $S = 18 - 17 + 13 - 11 + 9 - 6 + 1 = 7$. What is the sum of all

such alternating sums that are generated by the set
$N = \{1, 2, 3, \ldots, n\}$?

Counting the empty set, the number of subsets of a set of n elements is 2^n. Thus there are 2^n alternating sums S to be dealt with, 2^{n-1} which do not involve the integer n itself, and another 2^{n-1} which do contain n.

Clearly, a sum that contains n must start off with n in the leading position. In fact, corresponding to a sum S which contains n,

$$S = n - a + b - c + \cdots,$$

there is the sum $S' = a - b + c - \cdots$, containing all the same integers except n.

Of course, all the signs are reversed in S', with the result that

$$S + S' = n.$$

Since the sums in each of the 2^{n-1} pairs (S, S') add to n, the required grand sum of all the alternating sums must be just $n \cdot 2^{n-1}$.

5. U.S.A. Olympiad, 1983 (p. 173)

Prove that, in any *open* interval of length $1/n$ on the real number line, the number of irreducible fractions p/q which have denominators in the range $(1, 2, 3, \ldots, n)$ cannot exceed $(n + 1)/2$.

The first thing to observe is that a fraction may be given by more than one expression p/q even though q is restricted to the range $\{1, 2, \ldots, n\}$; for example, for $n = 23$, the fraction 3/5 would arise again as 6/10, 9/15, and 12/20. Since we are interested only in different *values*, we should include any fraction only once in our count, whatever pseudonyms it may have. Under which name shall we include it, then? Although its irreducible form 3/5 is the obvious choice, I suggest that we pick either 9/15 or 12/20.

The key to our solution is the observation that the first "half" of the range $\{1, 2, \ldots, n\}$ can be dispensed with completely. If a fraction

p/q has a denominator from the subrange,

$$A = \left\{1, 2, \ldots, \left[\frac{n}{2}\right]\right\},$$

then at least one of its equivalent forms

$$\frac{2p}{2q} = \frac{3p}{3q} = \frac{4p}{4q} = \cdots = \frac{kp}{kq} = \cdots$$

will have a denominator kq that lies in the upper subrange

$$B = \left\{\left[\frac{n}{2}\right] + 1, \left[\frac{n}{2}\right] + 2, \ldots, n\right\}.$$

(For $n = 23$, $A = \{1, 2, \ldots, 11\}$, $B = \{12, 13, \ldots, 23\}$.) Clearly we have

$$\text{length of } B = n - \left[\frac{n}{2}\right] = \frac{n}{2} + \frac{n}{2} - \left[\frac{n}{2}\right] \geq \frac{n}{2} \geq \left[\frac{n}{2}\right]$$
$$= \text{length of } A;$$

therefore the multiples of any denominator q in A, namely $2q, 3q, \ldots$, proceeding along in steps of size q (which is $\leq$ length of A), must find B too big a stretch to get over in a single stride, and will have at least one value in B itself. Thus every fraction we wish to count has an expression p/q, where $q \in B$, and so let us agree to count it under any one of its upper range names.

It remains only to observe that no two of our fractions can be counted under the same denominator in B. If $q \in B$ (actually A or B), then even the consecutive fractions p/q and $(p+1)/q$ differ by $1/q$, which is at least $1/n$, making the fractions too far apart for both of them to lie in an *open* interval of length $1/n$.

Thus each fraction enters our count with a *different* denominator from B, and so the required number N of different fractions must be $\leq |B|$. But, whether n is odd or even, it is easy to verify that $|B| = \left[\frac{n+1}{2}\right]$. For $n = 2k$

$$|B| = n - \left[\frac{n}{2}\right] = 2k - k = k,$$

and

$$\left[\frac{n+1}{2}\right] = \left[\frac{2k+1}{2}\right] = k;$$

for $n = 2k + 1$:

$$|B| = 2k + 1 - k = k + 1, \quad \text{and} \quad \left[\frac{n+1}{2}\right] = k + 1.$$

Hence we have $N \leq [\frac{n+1}{2}] \leq \frac{n+1}{2}$, as desired.

6. Problem P376 from the Hungarian Journal for High Schools (p. 238)

For a year or so, the following problem completely stumped me every time I looked at it.

> Determine the radii of all circles each of whose interior points can be colored either red or blue so that every pair of points in the circle that are 1 unit apart have different colors.

Of course the radius r of such a circle would have to be $> 1/2$ in order to give substance to the problem, and if the circle were big enough to contain an equilateral triangle of side 1, i.e., $r > 1/\sqrt{3}$, then the coloring would be impossible since at least two of the vertices would have to be the same color. Thus I had made only the barest start on things when I casually mentioned the problem to my colleague Ken Davidson. I had hardly stated the problem with the above prefatorial remarks about the equilateral triangle when it had already dawned on Ken that this notion could be extended to regular star-polygons with a large odd number of unit sides to show that there must always be two points, one unit apart, which are the same color whenever $r > 1/2$. The answer to the question, then, is that there is no such circle.

Of course Ken left it to me to work out the details (these bright guys always do), with the following result.

Solution by Ken Davidson. We begin with a given circle $O(r)$ of radius $r > 1/2$. Now, a regular polygon G can be inscribed in the circle to have any odd number of sides $2k + 1$. Because the number of sides is odd, each vertex A is opposite a *pair* of vertices which determine the opposite side YZ, and it is such "longest" diagonals AY and AZ that we are interested in. The greater $2k + 1$ is taken, the smaller YZ will get; by taking k large enough, we can make YZ so small that AY will be so close to being a diameter (of length $2r$ which is bigger than 1) that

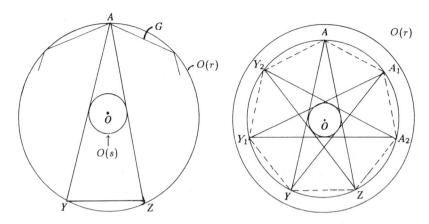

FIGURE 71

the length of AY will also exceed 1. Suppose, then, that k is chosen big enough to make $AY > 1$.

Clearly all such longest diagonals AY are tangents to a small circle $O(s)$ in the center of $O(r)$. Now, it is vital to our solution that the length AY of these tangents be exactly 1 unit. Therefore, let the figure be shrunk toward the center O in the ratio $1 : AY$; this will carry everything into the interior of the given circle, implying that all the image points will be colored, whereas the boundary of $O(r)$ was not colored to begin with. For simplicity, let's keep the same names for the images under this transformation, bearing in mind that now the length of every diagonal like AY is 1.

Now a tangent to $O(s)$ from a vertex of G meets the circumcircle in one of the two opposite vertices of G. Consequently, the sequence of tangents AY, YA_1, A_1Y_1, $Y_1A_2, \ldots$ carries one around G from A through the vertices $A, A_1, A_2, \ldots$, and after k such advances (through $2k$ tangents) we reach the opposite vertex Z in the form of A_k.

But all these "diagonal" tangents have length 1, and so the pairs of vertices they join must have different colors. Thus, if A is red, then Y is blue, A_1 is red, Y_1 blue, A_2 red, $\ldots$, to the conclusion that $A_k = Z$ is red. This means that both ends of the unit segment AZ are the same color, and our conclusion is established.

7. International Olympiad, 1983 (p. 207)

4. (revised).   If each point on the sides of equilateral triangle
ABC is colored either red or blue, prove that some 3 of the
points, which have the same color, are the vertices of a right-
angled triangle.

Since the given triangle is equilateral, presumably the solution will
hinge on some peculiarity of equilateral triangles. In view of the futility
of speculating at the outset on what this special property might be, let
us begin with some general observations.

In the given circumstances, the kind of right triangle that most
readily springs to mind is one like PQR which has a leg along a side of
ABC. Clearly, if P and Q are the same color, say red, then every point
R on BC, with the single exception of Q itself, would have to be blue in
order to avoid an all red triangle. Let's direct our efforts toward showing
that it is impossible to avoid creating a monochromatic right-triangle no
matter how the coloring might be carried out. So far we have that *if P*
and Q are both red, then the entire side BC must be blue, except for Q.

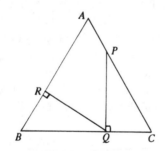 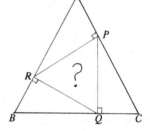

FIGURE 72

In this case, if any point S on BA or AC, other than B, P, or C, also
happened to be blue, then the perpendicular ST to BC would give an
all blue triangle SBT. Thus, if any pair of points like P and Q are both
red, all the rest of BC must be blue, and except for B and C themselves,
all of BA and AC must be red. But then, the perpendicular PV to AB
yields an all red triangle. We conclude, then, that *if* it is impossible to

avoid coloring some pair of points like P and Q the same color, then it is also impossible to avoid a monochromatic right-triangle.

Could it be that because ABC is equilateral, one can't help coloring some such pair P and Q the same color? Let's look into this a little further.

As shown, perpendiculars PQ and QR can generally be drawn to and from a point Q on a side of ABC. If either P and Q or Q and R were colored alike, a desired monochromatic pair would be produced. This is encouraging because the pigeonhole principle (you might have expected it would surface somewhere in the discussion) implies that some 2 of the 3 points P, Q, and R *must* be the same color. The trouble is, it could be P and R that are alike and Q the other color. From such a situation, we could only *be sure* of getting a monochromatic pair if, by some stroke of fortune, R and P were also an appropriate pair, that is, if RP and AC were also perpendicular.

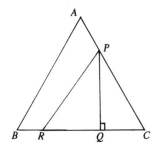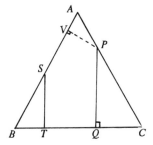

Perhaps this is where the equilateral character of triangle ABC comes into play. Thus it might occur to us to ask whether there exist 3 points P,Q,R on the sides of an equilateral triangle such that each side of $\triangle PQR$ is perpendicular to a side of ABC? Reaching the point of *asking* this question is the difficult part of the problem; answering it is relatively easy.

Since ABC is equilateral, $\angle C = 60°$, making PQC a 30-60-90 triangle and $PC = 2QC$. But the evident symmetry suggests $QC = AP$, in which case $PC = 2AP$, making P a point of trisection of AC. The obvious candidate, for our triangle PQR, is one determined by the appropriate 3 points of trisection of the sides.

Beginning with 3 such points of trisection, it is easy to see that the 3 little triangles PQC, ARP, and BQR are congruent (2 sides and the contained triangle); and since $\triangle PQC$ has a 60° angle at C and has $PC = 2QC$, it follows that $\angle PQC$, and similarly the angles at P and R, are right angles. The conclusion follows.

8. Problem M796 from Kvant (p. 269)

P is a point inside a square $ABCD$ such that $PA = 1$, $PB = 2$, and $PC = 3$. How big is $\angle APB$?

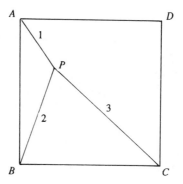

FIGURE 74

The following brilliant solution is by Murray Klamkin. Without prior experience in translating and rotating figures, it is unlikely, though not inconceivable, that one would think of spinning the figure through a right angle about the vertex B to a position $A'BAD'$, as shown (the first person to do such things must have been very clever, indeed). The segment BP is carried into BP' to form an isosceles right-triangle $P'BP$; this makes $\angle P'PB = 45°$, and we have a start on the desired angle APB.

By the Pythagorean theorem, we have $P'P = \sqrt{8}$ and, noting that $P'A = PC = 3$, we have

$$AP^2 + P'P^2 = 1 + 8 = 3^2 = P'A^2,$$

implying $\angle APP'$ is a right angle by the *converse* of the Pythagorean theorem. Hence

$$\angle APB = 90° + 45° = 135°.$$

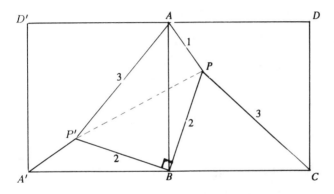

FIGURE 75

9. Problems from the Leningrad High School Olympiad, 1980 (pp. 302–305)

(a) Third Round, Grade 8

If a,b,c,d are positive real numbers that add up to 1, prove that

$$S = \sqrt{4a + 1} + \sqrt{4b + 1} + \sqrt{4c + 1} + \sqrt{4d + 1} < 6.$$

Solution by Mark Kantrowitz, Brookline, Massachusetts. By the arithmetic mean-geometric mean inequality we have

$$\frac{(4a + 1) + 1}{2} \geq \sqrt{(4a + 1) \cdot 1},$$

i.e.,

$$\sqrt{4a + 1} \leq 2a + 1,$$

with equality iff $4a+1 = 1$, i.e., iff $a = 0$; since a is positive, the inequality is strict, and we have

$$\sqrt{4a + 1} < 2a + 1.$$

Similar relations hold for b, c, and d, to give

$$S < (2a + 1) + (2b + 1) + (2c + 1) + (2d + 1)$$
$$= 2(a + b + c + d) + 4$$
$$= 2 + 4$$
$$= 6.$$

(b) Third Round, Grade 9

If, in $\triangle ABC$, angle A is twice angle B, prove that

$$a^2 = b(b + c),$$

where a is the length of the side opposite A, etc., as usual.

Let $\angle B = x$, making $\angle A = 2x$, and let K be the circumcircle of $\triangle ABC$. Let $AC(= b)$ be unfolded to AD in line with BA, and CD drawn. Then triangle ACD is isosceles with equal angles at C and D which add up to exterior angle $BAC = 2x$, making each of them x. Hence $\angle B = \angle D$, and we have $DC = BC = a$. Also, $\angle B = \angle DCA$, implying CD is a *tangent* to K. Finally, applying the basic theorem: *the product of a secant (DB) to a circle from a point (D) outside the circle* *and the part of it (DA) that lies outside the circle is equal to the square of the tangent (DC) from the same point,* we get

$$a^2 = DC^2 = DB \cdot DA = (b+c) \cdot b.$$

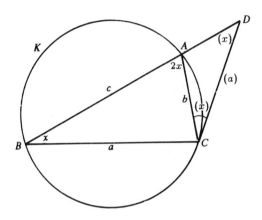

FIGURE 76

(c)   Third Round, Grade 9

If the magnitude of the quadratic function

$$f(x) = ax^2 + bx + c$$

never exceeds 1 for x in the closed unit interval $[0, 1]$, prove that the sum of the magnitudes of the coefficients cannot exceed 17:

$$|a| + |b| + |c| \leq 17.$$

Solution by Paul Wagner, Chicago, Illinois.   Giving x the values 0, 1/2, 1, we get, from $|f(x)| \leq 1$, that

$$|c| \leq 1, \tag{i}$$

$$\left|\frac{a}{4} + \frac{b}{2} + c\right| \leq 1, \tag{ii}$$

$$|a + b + c| \leq 1. \tag{iii}$$

Four times (ii) yields $|a + 2b + 4c| \leq 4$.

Since we have this information concerning the quantities

$$m = a + 2b + 4c, \quad \text{and} \quad n = a + b + c,$$

let's try to express the coefficients a, b, c in terms of these values. Straightforward eliminations immediately yield

$$a = -m + 2n + 2c, \quad \text{and} \quad b = m - n - 3c.$$

Hence

$$|a| \leq |m| + |2n| + |2c| \leq 4 + 2 + 2 = 8;$$
$$|b| \leq |m| + |n| + |3c| \leq 4 + 1 + 3 = 8.$$

Thus

$$|a| + |b| + |c| \leq 8 + 8 + 1 = 17.$$

(Note that $8x^2 - 8x + 1$ *does* satisfy $|f(x)| \leq 1$ for $x \in [0, 1]$.)

(d)   Third Round, Grade 10

How many different integers are there in the sequence

$$\left[\frac{1^2}{1980}\right], \left[\frac{2^2}{1980}\right], \cdots, \left[\frac{1980^2}{1980}\right];$$

where the square brackets denote the greatest integer function?

Solution by Daniel Ropp, Stillman Valley High School, Illinois. Since the first 44 terms in the sequence are all zero, there is obviously a lot of duplication in the first half of the sequence:

$$\left[\frac{1^2}{1980}\right], \ldots, \left[\frac{990^2}{1980}\right];$$

surprisingly, however, there is none in the second half! Despite all this duplication, we shall see that the first half realizes all 496 possible integers from 0 to $[990^2/1980] = 495$.

We proceed indirectly. Suppose that some integer $n < 495$ were to be missing from the first half of the sequence. Clearly the sequence is nondecreasing and, starting from 0, it works its way past n up to the value 495. Consequently, there must be a *last* integer $k < 990$ for which $[k^2/1980] < n$, after which the sequence immediately jumps over n to at least $n + 1$: $[(k+1)^2/1980] \geq n + 1$. For this k, then, we have

$$\frac{k^2}{1980} < n, \quad \text{and} \quad \frac{(k+1)^2}{1980} \geq n + 1,$$

i.e.,

$$\frac{k^2 + 2k + 1}{1980} \geq n + 1, \quad \text{and} \quad \frac{k^2}{1980} + \frac{2k+1}{1980} \geq n + 1;$$

recalling that $k^2/1980 < n$, this gives

$$\frac{2k+1}{1980} > 1, \quad 2k + 1 > 1980, \quad \text{and} \quad k \geq 990, \quad \text{a contradiction.}$$

Thus we can count on the first half for 496 different integers.

Now for the second half. Again we shall establish our claim by contradiction. Suppose, then, that some integer m were to occur twice in the second half. Since the sequence is nondecreasing, any repeated values must occur *consecutively* in the sequence and, for some $k > 990$, then, we would have

$$m = \left[\frac{k^2}{1980}\right] = \left[\frac{(k+1)^2}{1980}\right].$$

In this case,

$$m \leq \frac{k^2}{1980} < \frac{(k+1)^2}{1980} < m + 1,$$

implying

$$\frac{k^2 + 2k + 1}{1980} = \frac{k^2}{1980} + \frac{2k + 1}{1980} < m + 1,$$

where $k^2/1980 \geq m$. Hence

$$\frac{2k + 1}{1980} < 1, \quad \text{giving} \quad k < 990, \quad \text{a contradiction.}$$

Thus the first half generates 496 different integers and the second half 990 different integers. Before claiming a grand total of $496 + 990 = 1486$ different integers, however, we need to check that the last one in the first half doesn't occur again as the first one in the second half. But it is easily found that

$$\left[\frac{990^2}{1980}\right] = 495 \quad \text{and} \quad \left[\frac{991^2}{1980}\right] = 496,$$

giving the total 1486.

(e)   A Revised Problem

A, B, and C play a series of single table tennis games, starting with A and B in the first game, in which the loser drops out after each game.

(i)   If, at the end, A has won 10 games and B won 21 games, how many times did A and B play each other?

(ii)   If, at the end, A has won 10 games, B 15 games, and C 17 games, who lost the last game?

(i)   Each time a game is played, somebody sits out; thus, for each person, the total number $W + L + S$ of wins, losses and sit-outs will be the same, namely the total number of games in the series. Thus the situation is nicely summarized by a table of these values. If C won n games and x, y, z, respectively, denote their numbers of losses, we would have the following partial description of the results:

	W	L	S
A	10	x	
B	21	y	
C	n	z	

Now, in general, every time you lose, you sit out the next game. Thus losses and sit-outs generally pair up. However, there are two exceptions to this—C sat out the first game even though he hadn't lost the previous game, and the loser of the last game doesn't sit out after that because they quit at that point. It is not difficult to see, therefore, that C could *not* have lost the last game.

Suppose he did; then consider A: since A *didn't* lose the last game, his losses and sit-outs pair up exactly (he either sat out the last game because he lost the previous one—thus pairing them up, or he won the last game), giving him x sit-outs; similarly, B has y sit-outs; but then the totals for A and B, namely $10 + 2x$ and $21 + 2y$, couldn't possibly be equal since one is even and the other odd, a contradiction.

It must be, then, that either A or B lost the last game. Whoever it was, he got away without sitting out for this loss, implying $S = L - 1$. For the other one, we have $S = L$ since, without exception, every loss was followed by a sit-out.

Since C did not lose the last game, for him it must be that $S = L+1$: every loss *is* followed by a sit-out, plus the unprovoked sit-out at the beginning. Therefore we either have

	W	L	S
A	10	x	$x - 1$
B	21	y	y
C	n	z	$z + 1$

or

	W	L	S
A	10	x	x
B	21	y	$y - 1$
C	n	z	$z + 1$

Since the total number of games played is equal to the total number of losses, by equating the sum of the first two rows with twice the sum of the middle column, both these cases yield the same result

$$10 + 2x + 21 + 2y - 1 = 2(x + y + z),$$

$$2(x + y + 15) = 2(x + y + z),$$

giving $\quad z = 15 \quad$ and $\quad z + 1 = 16.$

Thus C sat out during the 16 times A and B played each other.

(ii) A similar analysis for this second part leads to

$$10 + 2x + 15 + 2y - 1 = 2(x + y + z),$$

giving $2(x + y + 12) = 2(x + y + z)$,   and $z = 12$.

As in part (i), C cannot have lost the last game. Now if A were to have lost the last game, then the first row of our table would be $A(10, x, x - 1)$, for a total of $2x + 9$, an odd number. However, from C's row we see that the total number of games played is $17 + 12 + 13 = 42$, an even number. Thus B must have lost the last game.

AN OFT-NEGLECTED FORM

As a and b run through the positive integers, the form $3ab + a + b$ obviously generates an infinite set of integers. Prove that there is also an infinity of positive integers none of which can be expressed in the form $3ab + a + b$.

Solution

It often happens that we respond almost involuntarily to the reading of a problem by conjuring up a few possibilities, or speculating on a general area where we might begin checking things out. But problems that ask you to prove that something can *not* be done leave you with little to grab hold of except the indirect approach. And, of course, it often works out that one *can* deduce a contradiction from the assumption that the thing in question can be done. For example, if the question were to show that no perfect square can be expressed in the form $4k + 2$, we would immediately deduce, in the event that some square should fulfill the condition, the existence of an impossible even square which is not divisible by 4. This works nicely because it concerns *all* squares.

But what do you do when the problem asks about a property that fails only in certain cases? In this event, a general contradiction is out of the question since the thing actually *does* work in some cases. For example, we might have been asked to show that there is an infinity of positive integers n which cannot be expressed in the form $3k$. It doesn't take long to see that one way around this particular problem is to show that there is another form which is incompatible with this one, say the form $3k + 1$. We need only show that none of the infinity of numbers of the form $3k + 1$ can be also expressed in the form $3k$. Therefore, in our present dilemma, we might consider the infinity of numbers of the form $3ab + a + b + 1$, and try to show that none of them can also be written in the form $3ab + a + b$.

The incompatibility of these forms is not a clear-cut issue like that of $3k$ and $3k + 1$, for it is conceivable that adjustments in a and b might permit a transformation from one form to the other. In fact, on my very first try I obtained the disappointing result:

$$3(6 \cdot 2) + 6 + 2 = 3(4 \cdot 3) + 4 + 3 + 1 \quad (= 44),$$

signalling the end to this particular line of thought. Of course, we could go on searching for a form that is incompatible with $3ab+a+b$. However, let us not abandon the indirect approach just yet. Even though a general contradiction is out of the question, perhaps an indirect approach might still turn up some useful information.

Suppose that a positive integer n were to be expressible in the form $3ab + a + b$:

$$n = 3ab + a + b.$$

Although noncompressible equations like this yield information grudgingly, we might eventually remember what we did earlier in a similar situation. When we were trying to show that the only solution in positive integers to the equation $ab = a+b$ is (2,2), we proceeded brilliantly to manipulate the equation until it presented us with an expression that could be factored:

$$ab - a - b + 1 = 1, \quad (a - 1)(b - 1) = 1, \quad \text{etc.}$$

Unfortunately these things never carry over from one application to another in a completely automatic way. In the present case, we need first to multiply through by 3 before adding 1:

$$3n + 1 = 9ab + 3a + 3b + 1$$
$$3n + 1 = (3a + 1)(3b + 1).$$

We have discovered, then, that whenever n can be expressed in the form $3ab + a + b$, the number $3n + 1$ can be factored into two integers, each at least 4. Well then, if $3n + 1$ were *not* to have such factors, as in the case of a prime, the number n couldn't possibly be of the form $3ab + a + b$.

The only question that remains is whether there exists an infinity of primes of the form $3n + 1$; and since Dirichlet's famous theorem confirms this, our solution is complete. (Dirichlet's theorem states that the arithmetic progression $\{an + b\}$ contains an infinity of prime numbers, provided a and b are relatively prime.)

A THEOREM OF LÉON ANNE

A point O in a quadrilateral $ABCD$, which is not a parallelogram, is joined to each vertex. If O moves so that each pair of opposite triangles at O contains half the area of $ABCD$, prove that O is constrained to lie on the line joining the midpoints of the diagonals.

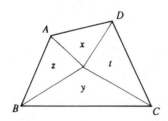 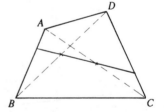

FIGURE 77

Solution

There is no shortage of proofs of this theorem. The following proof, due to Basil Rennie (James Cook University of North Queensland, Australia), surely is one of the most beautiful and elegant.

174

Let $O(x, y)$ be an arbitrary point on the locus, relative to any conve-
nient cartesian frame of reference. Now, the area of a triangle which has
two fixed vertices and only one variable vertex, (x, y), is a linear function
of x and y. Thus the condition governing the locus of O, involving only
the equality of sums of linear expressions, is given by a *linear* equation.
Accordingly, the locus must be confined to some straight line. And since
a median bisects the area of a triangle, the midpoints of the diagonals of
$ABCD$ are clearly seen to lie on the locus, and the conclusion follows.

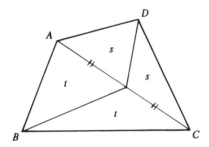

FIGURE 78

ᴄIAL PAIRS OF POSITIVE INTEGERS

Besides the trivial pairs (2,2) and (2,2), find two pairs of positive integers such that the sum of either pair is the product of the other.

Solution

If the pairs are (x, y) and (z, w), we require both

$$x + y = zw \quad \text{and} \quad z + w = xy.$$

Now what? Frankly, I never know what to do with things like this; even though someone may show you his solution to such a problem, he seldom really explains how he came to think of it. I guess that helps make mathematics an adventure; generally the price of enlightenment is, at the very least, the willingness to invest a certain amount of time and energy trying out various possibilities that somehow come your way.

Anyhow . . ., after playing with our equations for awhile, you might have rearranged them in the form $x + y - zw = 0$ and $z + w - xy = 0$, and put them together to obtain

$$x + y - zw + z + w - xy = 0.$$

At this point, you might reasonably wonder about factoring the left side, and thus be led to the crucial discovery that, by changing all the signs, adding a couple of 1's, and grouping related terms, the sections do factor to give

$$(x-1)(y-1) + (z-1)(w-1) = 2.$$

Since the variables are all positive integers, neither term here can be negative. The only way both can be *positive* and add up to 2, however, is to have each equal to 1, and this requires all the factors involved to be 1, giving

$$x = y = z = w = 2, \quad \text{the trivial solution.}$$

The only other possibility, then, is for one of the terms to be 2 and the other 0. A term equal to 2 can arise only from the factors 1 and 2, making the corresponding pair of integers 2 and 3. Since this pair has sum 5, then the other pair must be two factors of 5, namely (1,5), and we have the only other solution

$$(2,3) \quad \text{and} \quad (1,5).$$

AN INTRIGUING SEQUENCE

Determine the limit of the sequence $\{P_n\}$ which is defined by

$$P_1 = 4, \quad \text{and, for} \quad n = 1, 2, 3, \ldots,$$

$$P_{n+1} = 2^{n+1}\sqrt{2} \cdot \sqrt{1 - \sqrt{1 - \left(\frac{P_n}{2^{n+1}}\right)^2}}.$$

Solution

At first glance those nested square roots seem to imply terrible complications. However, they will be our salvation yet, for the inner root has the form $\sqrt{1 - y^2}$, which can hardly fail to recall the familiar relation $\sqrt{1 - \sin^2 x} = \cos x$. Of course, if $P_n/2^{n+1} > 1$, then it won't be the sine of any angle. Although this may come back to haunt us, let's set aside this unhappy possibility for the moment and see what would happen if we get lucky. In this case we would have

$$P_n = 2^{n+1} \sin x,$$

and from the given recurrence relation we get

$$P_{n+1} = 2^{n+1}\sqrt{2}\sqrt{1 - \sqrt{1 - \sin^2 x}}$$

$$= 2^{n+1}\sqrt{2}\sqrt{1 - \cos x}$$

$$= 2^{n+1}\sqrt{2}\sqrt{1 - \left(1 - 2\sin^2(x/2)\right)}$$
$$= 2^{n+1}\sqrt{2}\sqrt{2\sin^2(x/2)},$$

i.e.,

$$P_{n+1} = 2^{n+2}\sin\frac{x}{2}.$$

That is to say,

$$\text{if} \quad P_n = 2^{n+1}\sin x, \quad \text{then} \quad P_{n+1} = 2^{n+2}\sin\frac{x}{2}.$$

Consequently, if P_1 is able to assume this form, then they all can. Checking, we find that

$$2^2\sin x = P_1 = 4 \quad \text{implies} \quad \sin x = 1 \quad \text{and} \quad x = \frac{\pi}{2}.$$

Our luck has held, and it follows that

$$P_2 = 2^3\sin\frac{\pi}{2^2},$$
$$P_3 = 2^4\sin\frac{\pi}{2^3},$$
$$\cdots$$

and in general,

$$P_n = 2^{n+1}\sin\frac{\pi}{2^n}.$$

In the limit, then, we have

$$\lim_{n\to\infty} P_n = 2\pi \cdot \lim_{n\to\infty} \frac{\sin\frac{\pi}{2^n}}{\frac{\pi}{2^n}} = 2\pi \cdot 1 = 2\pi,$$

from the well known

$$\lim_{\theta\to 0+} \frac{\sin\theta}{\theta} = 1.$$

THAT NUMBER AGAIN

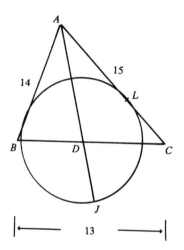

FIGURE 79

The lengths of the sides of triangle ABC are $AB = 14, BC = 13$, and $AC = 15$. The bisector of angle A meets BC at D. With center D, the circle is drawn that touches AC (since DA bisects angle A, this circle will also touch AB). Let the circle touch AC at L, and let AD

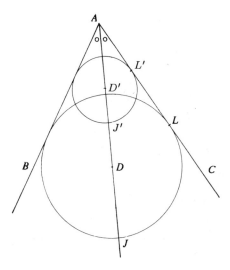

FIGURE 80

extended cross it at J. Prove that AJ/AL is none other than the golden ratio $(1 + \sqrt{5})/2$.

Let D' be any point on the bisector AD and let L' and J' be the points which correspond to L and J on the circle with center D' that touches AB and AC. Then the dilatation having center A and ratio AD/AD' carries the one circle into the other, in particular, L' and J' are carried, respectively, into L and J. That is to say, the ratio $AJ/AL = AJ'/AL'$ is the same for all circles that touch both arms of angle A, and can therefore be determined from the circle of our choice. In this case, let us abandon the given circle for the huge advantages that attend the *incircle* of $\triangle ABC$.

Let the pairs of equal tangents from the vertices have lengths x, y, z, as shown, and let the usual notations apply ($a = BC = y + z$, etc., the semiperimeter $s = a + b + c$, and so on). It is well known that the area Δ of the triangle ABC is given by both $\Delta = rs$ and the famous formula

$$\Delta = \sqrt{s(s - a)(s - b)(s - c)}$$

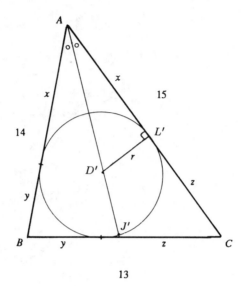

FIGURE 81

of the ancient Greek scholar Heron. Since $2s = 13 + 14 + 15 = 42$, we have $s = 21$, and it follows that

$$r = \frac{\Delta}{s} = \sqrt{\frac{(s-a)(s-b)(s-c)}{s}} = \sqrt{\frac{8 \cdot 6 \cdot 7}{21}} = 4.$$

We also have

$$21 = s = x + y + z = x + 13, \Rightarrow AL' = x = 8.$$

Then in right triangle $AD'L'$, the hypotenuse AD' is $\sqrt{80} = 4\sqrt{5}$; adding 4 for the radius $D'J'$ makes $AJ' = 4(\sqrt{5}+1)$. Finally, then,

$$\frac{AJ'}{AL'} = \frac{4(\sqrt{5}+1)}{8} = \frac{\sqrt{5}+1}{2},$$

the golden ratio!

A RATIONAL FUNCTION

For n a positive integer, does the fraction

$$f(n) = \frac{12n^3 - 5n^2 - 251n + 389}{6n^2 - 37n + 45}$$

ever reduce to an integer?

Solution

By long division we obtain

$$f(n) = 2n + 9 + \frac{15n^2 - 8n - 16}{6n^2 - 37n + 45}.$$

Thus $f(n)$ is an integer if and only if

$$g(n) = \frac{15n^2 - 8n - 16}{6n^2 - 37n + 45} \quad \text{is an integer.}$$

Now upon closer inspection, the parts of $g(n)$ factor into

$$g(n) = \frac{(3n - 4)(5n + 4)}{(3n - 5)(2n - 9)},$$

revealing the crucial fact that each part contains one of a pair of *consecutive* integers. Because consecutive integers are relatively prime, $g(n)$

183

can reduce to an integer only if

$$3n - 5 \mid 5n + 4.$$

In this case, $3n - 5$ will also divide the number

$$3(5n + 4) - 5(3n - 5),$$

which turns out to be just the integer 37. Consequently, $3n - 5$ must be one of the four divisors of 37, namely 1, -1, 37, -37, and we have, respectively, that $n = 2, \frac{4}{3}, 14, -\frac{32}{3}$. Thus the only positive integers n which can possibly make $g(n)$ an integer are 2 and 14.

But it takes more than $3n - 5$ dividing into $5n + 4$ to make $g(n)$ reduce to an integer. Checking out the other factors that are involved, we find that

$$g(2) = \frac{2 \cdot 14}{1(-5)} = -\frac{28}{5},$$

and

$$g(14) = \frac{38 \cdot 74}{37 \cdot 19} = 4.$$

Thus

$$f(14) = 41$$

is the only integral value taken by $f(n)$.

AN UNEXPECTED BIJECTION

One of the problems on an exam asked for the number of distinct 5-letter words that can be formed from the alphabet $\{a, a, a, b, b, b\}$. A student misread the question and worked out the number of such 6-letter words. But he got the right answer anyhow. Prove that the repetitions in the alphabet had nothing to do with it, that for any alphabet containing n letters, the number of $(n - 1)$-letter words is always the same as the number of n-letter words.

Solution

Clearly the entire alphabet is contained in each n-letter word and all but one letter of the alphabet occur in an $(n - 1)$-letter word. If the unused letter were to be stuck on the end of an $(n - 1)$-letter word, the result would be an n-letter word, and vice versa, if the last letter were to be snipped off an n-letter word, an $(n - 1)$-letter word would be left.

It is not difficult to see that such extensions and reductions determine a bijection between the sets of $(n - 1)$-letter and n-letter words. We need only show that

(i) no two extensions yield the same n-letter word, and

(ii) no two reductions leave the same $(n - 1)$-letter word.

(i)   Clearly, if you start with two different $(n - 1)$-letter words, you must wind up with different results no matter what you put on the end of them. Thus each $(n - 1)$-letter word corresponds to a different n- letter word.

(ii)   Conversely, suppose the last letters x and y are dropped from two different n-letter words p and q. If x and y are the same, then the difference between p and q must be due to different $(n - 1)$-letter first parts.

And if x and y are different, then the resulting $(n - 1)$-letter first parts of p and q still cannot be the same, because each of these first parts is composed from a different subalphabet (removing different things from the same total collection leaves different residues). Thus each n-letter word corresponds to a different $(n - 1)$-letter word, and the bijection is complete.

AN UNRULY SUM

The need to determine the sum of a series arises very frequently in mathematical investigations. In many cases, however, the series doesn't lend itself to easy evaluation and we have to settle for reasonable approximations.

Between what two integers does the sum S lie, where S is the unruly sum

$$S = \sum_{n=1}^{10^9} n^{-2/3} = 1 + \frac{1}{\sqrt[3]{2^2}} + \frac{1}{\sqrt[3]{3^2}} + \cdots + \frac{1}{\sqrt[3]{(10^9)^2}}?$$

Solution

Fermat and Descartes showed how much algebra can come to the aid of geometry; now geometry has a small opportunity to show its goodwill. Although the following general technique is very well known and widely used, this in no way detracts from its beauty and cleverness in the present instance.

It is evident from the figure that the terms of our sum are given by the areas of the shaded rectangles, making S equal to the total area that is shaded. Clearly, S amounts to more than the area under the curve $y = x^{-2/3}$ from $x = 1$ to $x = 10^9 + 1$, but is not as great as that under the curve $y = (x-1)^{-2/3}$. Thus we have

$$S > \int_1^{10^9+1} x^{-2/3}\, dx,$$

and we are tempted to claim similarly that

$$S < \int_1^{10^9+1} (x-1)^{-2/3}\, dx.$$

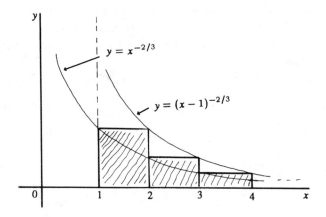

FIGURE 82

However, since $y = (x-1)^{-2/3}$ is *asymptotic* to the line $x = 1$, we are obliged to count the first rectangle separately and integrate from $x = 2$. Thus we have for the upper bound

$$1 + \int_2^{10^9+1} (x-1)^{-2/3}\, dx,$$

and the figure shows that this is indeed greater than S. Therefore

$$\int_1^{10^9+1} x^{-2/3}\, dx < S < 1 + \int_2^{10^9+1} (x-1)^{-2/3}\, dx,$$

$$3x^{1/3}\Big|_1^{10^9+1} < S < 1 + 3(x-1)^{1/3}\Big|_2^{10^9+1}$$

$$3\sqrt[3]{10^9+1} - 3 < S < 1 + 3000 - 3 = 2998.$$

Since $\sqrt[3]{10^9} < \sqrt[3]{10^9+1}$, we have

$$3\sqrt[3]{10^9} - 3 < S < 2998,$$

and finally

$$2997 < S < 2998.$$

GLEANINGS FROM MURRAY KLAMKIN'S OLYMPIAD CORNERS—1984

1. Brazilian Olympiad, 1983, (p. 40)

Show that the equation

$$\frac{1}{x} + \frac{1}{y} + \frac{1}{z} = \frac{1}{1983}$$

has a *finite number* of solutions in positive integers (x, y, z).

Since $1/x$, $1/y$, $1/z$ add up to $1/1983$, their average value is $(1/3) \cdot (1/1983)$, and since they can't all be below average, at least one of them, say $1/x$, must be at least $(1/3) \cdot (1/1983)$. Hence

$$\frac{1}{x} \geq \frac{1}{3(1983)}, \quad \text{and} \quad x \leq 3(1983).$$

On the other hand, if x were to be 1983 or less, then $1/x$ by itself would amount to the entire right side, or more, leaving no room for any positive quantities $1/y$ and $1/z$. Thus x must be at least 1984, giving

$$1984 \leq x \leq 3(1983), \quad \text{and} \quad \frac{1}{1984} \geq \frac{1}{x} \geq \frac{1}{3(1983)}.$$

As a result, then,

$$\frac{1}{y} + \frac{1}{z} = \frac{1}{1983} - \frac{1}{x} \geq \frac{1}{1983} - \frac{1}{1984} = \frac{1}{1983 \cdot 1984},$$

and arguing as above, one of $\frac{1}{y}, \frac{1}{z}$ must be at least $(1/2) \cdot 1/(1983 \cdot 1984)$, say $1/y$, and we have $y \leq 2(1983 \cdot 1984)$. Thus, with x and y each restricted to a finite range, there is only a finite number of pairs of values for x and y; and because z is determined once x and y are prescribed, there cannot be more solutions (x, y, z) than the finite number of pairs (x, y).

2. Michigan Mathematics Prize Competition, 1980, (p. 41)

Part of 5.   Consider the sequence

$$100, \ 55, \ 45, \ 10, \ 35.$$

The first two terms are prescribed, and thereafter each term is the difference between the preceding two terms, the later from the earlier. The sequence is terminated at 35 because the next term would have been negative; zeros are permitted, but not negative numbers. Had we begun with 100 and 60, a longer sequence would have resulted:

$$100, \ 60, \ 40, \ 20, \ 20, \ 0, \ 20.$$

Determine the integer B that leads to the longest sequence beginning

$$100, \ B, \ldots.$$

Solution by Murray Klamkin.   If the sequence is denoted by

$$a_1, \ a_2, \ldots, a_{n-2}, \ a_{n-1}, \ a_n, \ldots,$$

then the general rule of formation is $a_n = a_{n-2} - a_{n-1}$. With the simple rearrangement $a_{n-2} = a_{n-1} + a_n$, one might happen to notice that these

sequences are general "Fibonacci" sequences *when traced backwards.*
Thus, if x and y are the last two terms of a sequence, in reverse order
the sequence will simply be the Fibonacci sequence having initial terms
x and y:

$$x, \ y, \ x+y, \ x+2y, \ 2x+3y, \ 3x+5y, \ 5x+8y, \ 8x+13y, \ldots.$$

It is pretty obvious that the general term of this sequence is

$$t_n = [f(n-2)] \cdot x + [f(n-1)] \cdot y, \quad \text{for} \quad n > 2,$$

where $f(n)$ is the nth term of the original Fibonacci sequence

$$1, \ 1, \ 2, \ 3, \ 5, \ 8, \ 13, \ 31, \ 34, \ 55, \ 89, \ 144, \ 233, \ldots,$$

and the easy proof by induction is left to the reader. In passing, let us
note that the final term x of our original sequence couldn't be 0, but
that the penultimate term y could be:

$$x > 0, \quad y \geq 0.$$

Thus if y and x are the last two terms of a sequence of maximum
length that starts with the given positive integers 100 and B,

$$100, \ B, \ldots, y \ x,$$

100 would be the nth term t_n of the reversed sequence for which n *is
a maximum.* Accordingly, we can determine x and y as the appropriate
integral coefficients in

$$t_n = 100 = x \cdot f(n-2) + y \cdot f(n-1),$$

where n is as big as possible.

After that we can find B, which is just t_{n-1}, and actually generate the sequence. Since this equation concerns a pair of *consecutive* Fibonacci numbers, we need consider only the values

$$\{f(n-1), f(n-2)\} = \{(89, 55), \ (55, 34), \ (34, 21), \ldots\}.$$

Not forgetting that $y = 0$ is permissible, we soon find the first suitable integral combination to be

$$100 = 6(8) + 4(13),$$

making $x = 6$, $y = 4$, and $f(n - 2) = 8$, thus making $n - 2 = 6$, giving $n = 8$. Hence

$$\begin{aligned} B = t_7 &= x \cdot f(7 - 2) + y \cdot f(7 - 1) \\ &= 6 \cdot f(5) + 4 \cdot f(6) \\ &= 6(5) + 4(8) = 62. \end{aligned}$$

The longest sequence, then, is

$$100, \ 62, \ 38, \ 24, \ 14, \ 10, \ 4, \ 6,$$

ending in 4 and 6, as expected.

In Murray Klamkin's comprehensive solution (pages 10–12, 1985) to the general problem of finding the second term B for an arbitrary first term A, so as to yield a sequence of maximum length, (which was the second and major part of the problem of the Michigan Competition) he cleverly showed that

the best B is either $[lA]$ or $[lA] + 1$,

where $l = (\sqrt{5} - 1)/2 = .618$, the reciprocal of the golden ratio, and the square brackets indicate the greatest integer function.

Accordingly, for $A = 100$, this formula yields

the optimum $B = [61.8] = 61$, or 62.

Since $B = 61$ gives only 7 terms,

$$100, \ 61, \ 39, \ 22, \ 17, \ 5, \ 12,$$

the best B is 62 alright.

In closing we note that these two choices for B can lead to quite different results, and so one can't afford to be sloppy in making his choice:

for example, for $A = 99$: $B = [99(.618...)] = 61$, giving 99, 61, 38, 23, 15, 8, 7, 1, 6, with 9 terms, while $B = 62$ gives 99, 62, 37, 25, 12, 13, with only 6 terms.

3. Problem P383 from Kozepiskolai Matematikai Lapok (Hungarian Journal for High Schools) (p. 76)

Does there exist a multiple of 5^{100} which contains *no zero* in its decimal representation?

Solution by Andy Liu, University of Alberta. Sometimes it is much easier to solve a more general problem, seemingly more complicated and difficult, than to mount a direct attack on a problem in its given form. Craftily, Professor Liu shows, by induction, that it is very easy to prove for every positive integer k that there exists a k-*digit* multiple m_k of 5^k which does not contain a zero in its decimal representation.

For $k = 1$, clearly 5, itself, is the obvious and only choice. Suppose, then, for some $k \geq 1$, that m_k is a k-digit multiple of 5^k which contains no zero. Now, the desired m_{k+1} is only to have one more digit than m_k, and so it behooves us to investigate whether it is possible simply to extend m_k into m_{k+1} by attaching a nonzero digit at one end or the other. At the units end, only the nonzero digit 5 will give a multiple of 5, and this generally does not yield a multiple of 5^{k+1} (e.g., 5^2 does not divide 55). Therefore, let a nonzero digit t be added at the beginning of the k-digit m_k to give a tentative

$$m_{k+1} = 10^k t + m_k.$$

Since 5^k divides m_k, and also 10^k, then certainly $5^k \mid m_{k+1}$. Suppose $m_k = 5^k q$, giving $m_{k+1} = 5^k(2^k t + q)$. In order to make m_{k+1} divisible by 5^{k+1}, we need only have $2^k t + q$ provide a single factor 5. Clearly, the thing to do is to pick t so that $5 \mid 2^k t + q$. If $q \equiv 0 \pmod 5$, then $t = 0$ satisfies the condition, but in this case we must take the equivalent $t = 5$ in order to maintain nonzero digits. Since we are working here modulo 5, t can always be chosen from among the digits $\{1, 2, 3, 4, 5\}$.

From $m_1 = 5$, we can easily calculate the next few terms; although calculating m_{100} is out of the question, its existence is assured.

k	m_k	q	$2^k t + q \pmod 5$	t	m_{k+1}
1	5	1	$2t + 1$	2	25
2	25	1	$4t + 1$	1	125
3	125	1	$3t + 1$	3	3125
4	3125	5	$t + 5$	5	53125
5	53125	17	$2t + 2$	4	453125
6	453125	29	$4t + 4$	4	4453125

. . . .

4. Leningrad High School Olympiad, 1981, Grade 10, (p. 142)

Prove that the positive root of

$$x(x + 1)(x + 2) \cdots (x + 1981) = 1$$

is less than $1/1981!$.

In our efforts to solve an equation we are so accustomed to working toward the *isolation* of the variable that the following maneuver to a position in which the variable is very awkwardly given in terms of itself is hardly in the accepted class of useful manipulations and may well be rejected almost subconsciously as a retrograde step: for *positive x*, clearly

$$x = \frac{1}{(x + 1)(x + 2) \cdots (x + 1981)} < \frac{1}{1 \cdot 2 \cdots 1981} = \frac{1}{1981!}.$$

5. American Invitational Mathematics Examination, 1984, (pp. 143–4)

(i) 4. (revised). Let S be a collection of positive integers, not necessarily distinct, which contains the number 68. The average of the numbers in S is 56; however, if a 68 is removed, the average would drop to 55. What is the largest number that S can possibly contain?

Let $S = \{68, a_1, a_2, \cdots, a_{n-1}\}$. Then

$$\frac{1}{n} \cdot (68 + a_1 + a_2 \cdots + a_{n-1}) = 56,$$

i.e.,

$$68 + a_1 + a_2 + \cdots + a_{n-1} = 56n,$$

while

$$\frac{1}{n-1}(a_1+a_2+\cdots+a_{n-1}) = 55, \quad \text{or} \quad a_1+a_2+\cdots+a_{n-1} = 55(n-1).$$

Hence

$$68 + 55(n - 1) = 56n,$$

and we quickly have $n = 13$.

Then

$$68 + a_1 + a_2 + \cdots + a_{12} = 13(56) = 728,$$

giving

$$a_1 + a_2 + \cdots + a_{12} = 660.$$

If the largest member of S is $a_{12} = x$, then

$$x = 660 - (a_1 + a_2 + \cdots + a_{11}).$$

Now, for S to possess the greatest number x that it possibly can, all the others, which aren't the prescribed 68, must have the minimum value of 1, making S the collection $\{68, x, 1, 1, 1, \ldots\}$. In this case

$$x = 660 - 11 = 649,$$

and S is $\{649, 68, 1, 1, 1, 1, 1, 1, 1, 1, 1, 1, 1\}$.

(ii)   14.   What is the greatest even integer which *cannot* be written as the sum of two odd composite positive integers?

The first few odd composite positive integers are

$$9, \ 15, \ 21, \ 25, \ 27, \ 33, \ 35, \ 39, \ 45, \ 49, \ 51, \dots.$$

For an even integer $2n$, then, the possible sums in question are

$$9 + (2n - 9), \ 15 + (2n - 15), \ 21 + (2n - 21), \dots, \ k + (2n - k),$$

where k is the greatest odd composite integer $\le 2n - 9$ ($2n - k$ must be ≥ 9 in order to be composite, implying $2n - 9 \ge k$). We seek the greatest integer $2n$ for which all these terms in brackets are prime numbers.

It is easy to check small integers:

$30 = 9 + 21$;	32 can't be done;	$34 = 9 + 25$;
$36 = 9 + 27$;	38 can't be done;	$40 = 15 + 25$;
$42 = 9 + 33$;	$44 = 9 + 35$;	$46 = 21 + 25$;
$48 = 9 + 39$;	$50 = 15 + 35$;	$52 = 27 + 25$; ….

There is a chance that the answer is 38; it certainly fails in all cases:

$$(9, 29), \ (15, 23), \ (21, 17), \ (25, 13), \ (27, 11), \ (33, 5), \ (35, 3)$$

(recall that 1 is not a prime). In fact, it is not very difficult to see that 38 is indeed the maximum.

Suppose $2n$ ends in the digit d. Then for the 5 possible values of d we may proceed as follows: for

$d = 0$,	use	$2n = 15 + (2n - 15)$;
$d = 2$,	use	$2n = 27 + (2n - 27)$;
$d = 4$,	use	$2n = 9 + (2n - 9)$;
$d = 6$,	use	$2n = 21 + (2n - 21)$;
$d = 8$,	use	$2n = 33 + (2n - 33)$.

In each case the bracket is clearly odd, and because it ends in a 5 and is greater than 5 (in view of $2n > 38$), it must be composite. For example, we have $40 = 15 + 25$, $50 = 15 + 35$, and so on.

5. Canadian Olympiad, 1984, (p. 181)

Prove that, among any seven real numbers $y_1, y_2, \ldots, y_7$, some two, y_i and y_j, are such that

$$0 \le \frac{y_i - y_j}{1 + y_i y_j} \le \frac{1}{\sqrt{3}}.$$

Who would suspect that an extremely beautiful solution can be based on the pigeonhole principle?

Surely it is not farfetched to suggest that the form of the given expression might well bring to mind the formula for the tangent of the difference between two angles:

$$\tan(x_i - x_j) = \frac{\tan x_i - \tan x_j}{1 + \tan x_i \tan x_j}.$$

Thus, if $y_i = \tan x_i$ and $y_j = \tan x_j$, the expression in question would be just $\tan(x_i - x_j)$. Accordingly, we could reasonably be led to the transformation

$$y_n = \tan x_n, \quad n = 1, 2, \ldots, 7,$$

mapping the seven given real numbers $y_1, y_2, \ldots, y_7$, which could range anywhere between $-\infty$ and $+\infty$, to seven images $x_1, x_2, \ldots, x_7$ in the range of length π between $-\frac{\pi}{2}$ and $+\frac{\pi}{2}$.

By the pigeonhole principle, some two of the x's must differ by not more than $\frac{\pi}{6}$:

$$0 \le x_i - x_j \le \frac{\pi}{6};$$

and since the tangent function is strictly increasing in $(-\frac{\pi}{2}, \frac{\pi}{2})$, we have

$$\tan 0 \le \tan(x_i - x_j) \le \tan \frac{\pi}{6},$$

which is

$$0 \le \frac{y_i - y_j}{1 + y_i y_j} \le \frac{1}{\sqrt{3}}.$$

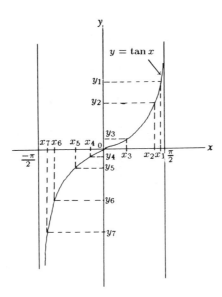

FIGURE 83

6. Swedish Olympiad, 1983, (p. 214)

5.   A unit square is to be covered by 3 congruent circular disks.

(a)   Show that there are disks of diameter less than the diagonal of the square that provide a covering.

(b)   Determine the smallest possible diameter.

(a)   Part (a) is so easy that one can't help wondering why it was asked. If EF is the midline between AB and CD in unit square $ABCD$, then clearly the circumcircles of the 2 halves cover the whole square and each has diameter equal to BE, which is obviously less than the diagonal BD.

It is somewhat puzzling that we have a disk left over, for only 2 of the permissible 3 have been used. However, the people who set these olympiads are very sharp characters and would be unlikely to ask a really easy question without some ulterior motive. Therefore it is a pretty good bet that some refinement of the above configuration is of major

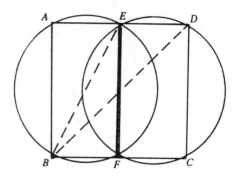

FIGURE 84

importance in determining the smallest possible size of 3 equal covering disks in part (b).

(b)   Perhaps it's just as well that this disguised hint was offered, for part (b) is a very difficult problem. Personally, I never had the chance of trying it on my own at the time of this olympiad for I had already written up Joe Lipman's 1958-solution to the problem in one of my "Mathematical Gems" columns in *The College Mathematics Journal* (then called *The Two-Year College Mathematics Journal*; see 1980, pages 116–7). It is such a nice solution that it is worth retelling here.

Obviously the circles of part (a) are far bigger than necessary when 3 of them are available, and so one might think of shrinking them down toward the chords BF and FC along the base of the square.

This would reduce the rectangular regions that they circumscribe and leave an essentially rectangular section $AXYD$ at the top for the third disk to cover. Since the 3 disks are to be the same size, this shrinking process could be continued until equal diameters $BE'(= E'C)$ and XD are attained. If $AX = x$ at this point, then

$$XD = \sqrt{x^2 + 1} = BE' = \sqrt{(1 - x)^2 + 1/4}$$

and we have $x^2 + 1 = 1 - 2x + x^2 + \frac{1}{4}$, giving $x = \frac{1}{8}$, and equal diameters of $\sqrt{65/64}$.

Of course, there is no guarantee that this is the desired minimum, but there is a chance that it might be and so some effort spent to this

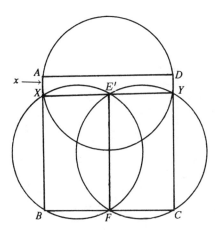

FIGURE 85

end would not be out of place at this point. From here, I suppose it is not inconceivable that one might proceed along the following brilliant line of thought.

We can imagine the disks being put down one at a time to partition the square into 3 disjoint subsets R, W, and B, for red, white, and blue; that is, let the points covered by the first disk be colored red, the additional points covered by the second disk white, and the remainder, covered by the third disk, blue.

Since each disk needs its full extent to cover its region, the diameter of each of the sets R, W and B is given by the diameter of its covering disk, $\sqrt{65/64}$. Now, if it were possible to cover the square with 3 equal disks of smaller size, they would similarly partition the square into 3 sets each of whose diameters would be less than $\sqrt{65/64}$. Joe Lipman shows that 3 such sets, *all* of smaller diameter, do not exist, implying that no 3 equal smaller covering disks could either.

His proof is by contradiction. Suppose that it *is* possible to color the square with 3 colors, red, white, and blue, so that the points of each color have diameter less than $\sqrt{65/64}$. Because there are four vertices and only 3 colors, the pigeonhole principle implies that two of the vertices must be the same color, say red. Now, a pair of opposite vertices must be done in different colors since their distance apart is $\sqrt{2}$, which exceeds

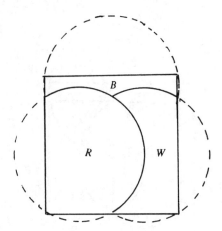

FIGURE 86

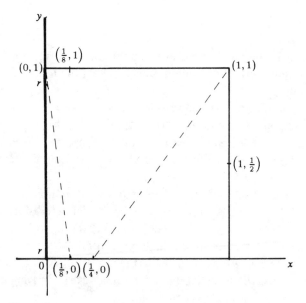

FIGURE 87

the diameter, which is even less than $\sqrt{65/64}$. Let coordinate axes be assigned to the square, as shown, and suppose that the adjacent vertices $(0,0)$ and $(0,1)$ are the red ones.

Since each of the five points (1/8,0), (1/8,1), (1/4,0), (1,1/2), (1,1) is at least $\sqrt{65/64}$ from one or both of the red vertices (0,0) and (0,1), none of them can also be red. The three points (1/8,0), (1/8,1), (1,1/2), among these five, then, must be either white or blue, and the pigeonhole principle implies that two of them have the same color, say white. However, (1,1/2) is farther than $\sqrt{65/64}$ from either of the other two, making (1/8,0) and (1/8,1) the two white ones.

As far as the remaining two of the five white or blue points is concerned, we have (1/4,0) too far from (1/8,1) to be white, and (1,1) too far from (1/8,0) to be white. Thus *they must both be blue*, implying the contradiction that the blue set has diameter at least equal to the distance 5/4 between them, which far exceeds $\sqrt{65/64}$.

7. The Mathematical Balkaniad, 1984, (p. 310)

Let $x_1, x_2, \cdots, x_n$, where $n \geq 2$, be positive numbers that add up to 1. Prove that

$$S = \frac{x_1}{1 + x_2 + x_3 + \cdots + x_n} + \frac{x_2}{1 + x_1 + x_3 + \cdots + x_n}$$

$$+ \cdots + \frac{x_n}{1 + x_1 + \cdots + x_{n-1}}$$

$$\geq \frac{n}{2n - 1}.$$

Solution by Murray Klamkin. Since the x's add up to 1, the denominators in S are simply $2 - x_i$, and

$$S = \sum_{i=1}^{n} \frac{x_i}{2 - x_i}.$$

Now Murray very craftily manipulates the general term as follows:

$$\frac{x_i}{2 - x_i} = \frac{2 - 2 + x_i}{2 - x_i} = \frac{2}{2 - x_i} - 1,$$

to get

$$S = \sum_{i=1}^{n} \frac{x_i}{2 - x_i} = \left[2 \sum_{i=1}^{n} \frac{1}{2 - x_i} \right] - n.$$

Following a neat application of the Cauchy inequality

$$\Big((a_1^2 + a_2^2 + \cdots + a_n^2)(b_1^2 + b_2^2 + \cdots + b_n^2) \geq (a_1 b_1 + a_2 b_2 + \cdots + a_n b_n)^2\Big)$$

with $a_i = 1/\sqrt{2 - x_i}$, and $b_i = \sqrt{2 - x_i}$, he gets

$$\Big[\sum_{i=1}^{n} \frac{1}{2 - x_i} \Big] \big[(2 - x_1) + (2 - x_2) + \cdots + (2 - x_n) \big] \geq n^2,$$

and

$$\sum_{i=1}^{n} \frac{1}{2 - x_i} \geq \frac{n^2}{2n - 1}.$$

Hence

$$\begin{aligned} S &= 2 \sum_{i=1}^{n} \frac{1}{2 - x_i} - n \\ &\geq \frac{2n^2}{2n - 1} - n \\ &= \frac{n}{2n - 1}, \end{aligned}$$

as required.

AN INTERESTING INEQUALITY

If the positive real numbers $x_1, x_2, \ldots, x_{n+1}$ are such that

$$\frac{1}{1 + x_1} + \frac{1}{1 + x_2} + \cdots + \frac{1}{1 + x_{n+1}} = 1,$$

prove that

$$x_1 x_2 \cdots x_{n+1} \geq n^{n+1}.$$

Solution

If we set $1/(1 + x_i) = a_i$, then we have

$$a_i + a_i x_i = 1 \quad \text{and} \quad x_i = \frac{1 - a_i}{a_i}.$$

Denoting $a_1 a_2 \cdots a_{n+1}$ by A, we have the product

$$x_1 x_2 \cdots x_{n+1} = \prod_{i=1}^{n+1} \frac{1 - a_i}{a_i} = \frac{\prod_{i=1}^{n+1}(1 - a_i)}{A}.$$

Now the given condition is

$$a_1 + a_2 + \cdots + a_{n+1} = 1,$$

and so $1 - a_i$ is just the sum of the n a's other than a_i:

$$1 - a_i = a_1 + \cdots + a_{i-1} + a_{i+1} + \cdots + a_{n+1}.$$

Applying the arithmetic-geometric mean inequality to these n other a's, we get, for each i, that

$$\frac{a_1 + \cdots + a_{i-1} + a_{i+1} + \cdots + a_{n+1}}{n} \geq [a_1 \cdots a_{i-1} a_{a+1} \cdots a_{n+1}]^{1/n}.$$

That is,

$$\frac{1 - a_i}{n} \geq \left[\frac{A}{a_i}\right]^{1/n}.$$

Multiplying the $n + 1$ such relations gives

$$\frac{\prod_{i=1}^{n+1}(1 - a_i)}{n^{n+1}} \geq \left[\frac{A^{n+1}}{a_1 a_2 \cdots a_{n+1}}\right]^{1/n} = \left[\frac{A^{n+1}}{A}\right]^{1/n} = A.$$

Thus

$$x_1 x_2 \cdots x_{n+1} = \frac{\prod_{i=1}^{n+1}(1 - a_i)}{A} \geq n^{n+1},$$

as desired.

A SERIES OF RECIPROCALS

In former days the topic of double-arithmetic series was a popular sup-plementary topic in elementary algebra courses. In the present morsel we are required to add up a series of *reciprocals* of such terms: deter-mine

$$S = \sum_{i=1}^{\infty} \frac{1}{i(k+i)} = \frac{1}{1(k+1)} + \frac{1}{2(k+2)} + \cdots,$$

where k is a given positive integer.

Solution

A series can sometimes be expressed as a "telescoping sum of differ-ences." If each term can be written as a difference between two quan-tities a_i and b_i which have the *same form*, with the additional property that the negative part of each term cancels the positive part of the next term (i.e., $b_i = a_{i+1}$), then the partial sums telescope to the difference between just two terms:

$$S_n = (a_1 - b_1) + (a_2 - b_2) + \cdots + (a_n - b_n) = a_1 - b_n.$$

In the hope of such a compression, we might begin by resolving our general term into its partial fractions. Setting

$$\frac{1}{i(k+i)} = \frac{A}{i} - \frac{B}{k+i} = \frac{(A-B)i + Ak}{i(k+i)},$$

we obtain $A - B = 0$ and $Ak = 1$, giving $A = B = 1/k$. Thus, while the general term is not exactly a difference of like expressions, it is a constant $1/k$ times a difference of quantities of the same basic form:

$$\frac{1}{i(k+i)} = \frac{1}{k}\left(\frac{1}{i} - \frac{1}{k+i}\right).$$

The constant factor $1/k$ gives no trouble since it can be factored out of every term. Although the negative part of one difference doesn't cancel the positive part of the very next difference, it does nullify the positive part of the difference k terms farther along the series. Skipping the intervening terms doesn't really matter; this simply means that the positive parts of the first k differences don't get cancelled, that's all. Hence the required sum is the remnant

$$S = \frac{1}{k}\left(1 + \frac{1}{2} + \frac{1}{3} + \cdots + \frac{1}{k}\right).$$

ON THE LEAST COMMON MULTIPLE

Let x and y be positive integers, and let (x,y) and $[x,y]$, respectively, denote their greatest common divisor and least common multiple. Since (x,y) seems to arise so much more often than $[x,y]$, perhaps you might enjoy the following little problem featuring $[x,y]$.

The sum and least common multiple of two positive integers x, y are given: $x + y = 5432$, and $[x,y] = 223020$; find the numbers.

Solution

Perhaps the first thing to come to mind is the basic relation

$$x \cdot y = [x,y] \cdot (x,y).$$

(Certainly xy is a common multiple of x and y, and is too much to be $[x,y]$ by precisely the factor (x,y), which is contained in x and repeated in y.) We already know $x+y$, and if we could determine the value of xy, we could easily solve for x and y, themselves. We lack only the value of (x,y) in order to get xy from the relation above.

Eventually, one might wonder whether (x,y) could possibly be just the greatest common divisor of the two given quantities: that is, is

$$(x,y) = (x + y, [x,y])?$$

209

To check this out, the standard representations are useful: let

$$(x, y) = d, \qquad \frac{x}{d} = a, \qquad \frac{y}{d} = b;$$

then a and b are relatively prime, i.e., $(a, b) = 1$. In these terms,

$$[x, y] = abd,$$

and we have

$$(x + y, [x, y]) = ((a + b)d, abd) = d \cdot (a + b, ab).$$

We would like $(a + b, ab) = 1$. On the contrary assumption that $(a + b, ab) = k > 1$, any prime divisor p of k would have to divide either a or b since k divides ab, and because p likewise divides $a + b$, it would have to divide the other as well, contradicting $(a, b) = 1$. Hence $(a + b, ab) = 1$ alright, and we have

$$(x, y) = (x + y, [x, y]) = (5432, 223020) = 28$$

(by the Euclidean algorithm). Thus

$$x + y = 5432, \quad \text{and} \quad xy = (223020) \cdot 28.$$

Although the solution of these equations is perfectly straightforward, we might as well take advantage of the knowledge that $(x, y) = 28$, which implies that both equations are divisible by 28. Thus we have

$$x + y = ad + bd = a(28) + b(28) = 5432,$$

and

$$xy = ab(28^2) = (223020) \cdot 28,$$

giving the simpler

$$a + b = 194, \quad \text{and} \quad ab = 7965.$$

Thus a and b are the roots of the quadratic equation

$$z^2 - 194z + 7965 = 0,$$

and we obtain $a = 59$, $b = 135$, and $x = 28 \cdot 59$, $y = 28 \cdot 135$, i.e.,

$$x = 1652, \qquad y = 3780.$$

A FAMILY OF EQUATIONS

Show that the equation

$$1 + 2x + 3x^2 + \cdots + nx^{n-1} = n^2$$

has a rational root between 1 and 2 for all $n = 2, 3, 4, \ldots$.

Solution

(a) To a student of the 1990's, the series

$$S = 1 + 2x + 3x^2 + \cdots + nx^{n-1}$$

might be considered to be unsummable. In the good old days, summing an arithmetic-geometric series was sometimes included in a standard course. The key move consists in multiplying through by the common ratio of the geometric series:

$$S = 1 + 2x + 3x^2 + \cdots + \qquad nx^{n-1},$$
$$xS = \qquad x + 2x^2 + \cdots + (n-1)x^{n-1} + nx^n.$$

Subtracting, we get

$$(1-x)S = 1 + x + x^2 + \cdots + x^{n-1} - nx^n$$
$$= \frac{1-x^n}{1-x} - nx^n,$$

giving

$$S = \frac{1-x^n}{(1-x)^2} - \frac{nx^n}{1-x}$$
$$= \frac{1-x^n - nx^n(1-x)}{(1-x)^2}$$
$$= \frac{1-x^n[1+n(1-x)]}{(1-x)^2} \quad \text{(for } x \neq 1, \text{ of course)}.$$

We are trying to show that the equation $S = n^2$ has a root $x = 1+y$, where y is some positive rational number < 1. In these terms, we would like to have $1 - x = -y$, and

$$S = \frac{1-(1+y)^n[1-ny]}{(-y)^2} = \frac{1}{y^2} - \frac{(1+y)^n(1-ny)}{y^2} = n^2.$$

One conceivable way of achieving the desired result would be to have the term $1/y^2$ equal to n^2 and have the rest of S vanish altogether; perhaps this is too much to hope for, but mathematicians learned long ago that, despite the odds, it often pays to guess boldly and check things out.

For $1/y^2 = n^2$, then, we would have $y = 1/n$, which is a positive rational number alright, and furthermore, the awkward second part of S does indeed vanish in view of its zero factor $(1-ny)$. Thus $x = 1+(1/n)$ always satisfies the equation, and when this fact is stated in the form

$$1 + 2\left(1+\frac{1}{n}\right) + 3\left(1+\frac{1}{n}\right)^2 + \cdots + n\left(1+\frac{1}{n}\right)^{n-1} = n^2,$$

we obtain a very interesting identity.

(b)   While the above approach might be a fairly routine procedure for those of us who have been studying a little longer, the problem is not beyond a present-day student; however, he would have to show a certain amount of initiative.

As teachers, we try never to miss an opportunity to undermine the divisive effect of the artificial barriers that exist in the mind of the student between the various branches of mathematics. These partitions are an unavoidabe consequence of the fact that our vast chain-like subject simply has to be taught one topic at a time in the lower grades. At the higher levels, however, it is always a special delight to experience the deeper unity of mathematics by bringing together tools from several areas in the solution of a difficult problem.

A keen modern student might notice that the series S is just the derivative of the polynomial

$$P = x + x^2 + x^3 + \cdots + x^n.$$

Consequently we can derive a formula for S by *first* adding the series P and *then* differentiating:

$$
\begin{aligned}
S &= \frac{dP}{dx} = \frac{d}{dx}\left[\frac{x - x^{n+1}}{1 - x}\right] \\
&= \frac{(1 - x)[1 - (n + 1)x^n] - (x - x^{n+1})(-1)}{(1 - x)^2},
\end{aligned}
$$

which simplifies easily to the formula found above.

Gleanings from Murray Klamkin's Olympiad Corners—1985

1. Four Problems from Olympiads in the U.S.S.R., 1984 (pp. 2–5)

(a)   Leningrad Olympiad

6.   Two players, A and B, take turns, A first, inserting their choice of one of the signs $+$, $-$, or $\times$ in any of the empy places between the terms

$$1 \quad 2 \quad 3 \quad 4 \quad \cdots \quad 99 \quad 100$$

until an arithmetic expression S is determined at the end of 99 turns. Prove that A can make S turn out to be odd or even as he wishes.

Solution by Murray Klamkin.   Since we are concerned only with the parities of quantities and not their sizes, a $+$ sign and a $-$ sign are equivalent, and so let us dispense with the $-$ sign and always use $+$ as the alternative to the $\times$ sign. Furthermore, because it is only parities that count, the initial set-up is adequately described by the alternating sequence of O's and E's, for odd and even,

$$O \quad E \quad O \quad E \quad \cdots \quad O \quad E$$

instead of 1 2 3 ⋯ 100. Also, we should note that, in evaluating the final expression S, the multiplications are performed before the additions.

Thus one can't always be certain of the effect of a turn until the neighboring signs have been entered. For example, it appears that the $+$ in $\cdots E + O \cdots$ unites an odd and an even term to give an odd result; however, if a $\times$ sign is entered on a later turn in the next position to give $\cdots E + O \times E \cdots$, then the $+$ winds up between two even values. Finally, let us observe that a $\times$ sign, inserted between an E and O in either order, $E \times O$ or $O \times E$, gives a result that is unalterably even and can therefore be replaced by a single E. Now then, let's see how A can make S turn out to be even.

Clearly A not only has an extra turn to spend but, being the first to play, he can take advantage of whatever opportunities exist to set things up to suit his purpose. A can make all the components of S turn out to be even numbers if he sticks to using $\times$ signs exclusively, puts the first one in the first space to give

$$O \times E \ \ O \ E \cdots O \ E = E \ O \ E \ O \ \ E \cdots E \ \ O \ E,$$

in which every O is sandwiched between two E's, and thereafter simply plays his $\times$ on the *other side* of the O that B plays next to. B can't help playing his choice of sign, $\star$, beside some O, to give either

$$E \star O \ E \cdots \quad \text{or} \quad E \ O \star E \cdots,$$

to which A replies

$$\cdots E \star O \times E \cdots \quad \text{or} \quad E \times O \star E \cdots,$$

resulting in

$$\cdots E \star E \cdots \quad \text{or} \quad E \star E \cdots.$$

Thus, on each turn A neutralizes an odd quantity, and eventually all the odd numbers are effectively eliminated, leaving S unavoidably even.

To make S odd, A simply starts off with a $+$ in the first position to give

$$O + \underbrace{E \ O \ E \cdots O \ E}_{Y},$$

and thereafter plays as above; as in the earlier case, this results in nothing but even values in the large expression Y after the $+$ sign, eventually making

$$S = O + E,$$

which is odd.

(b)   Tournament of Towns

9.   The 17 boys and 17 girls in a ballroom dancing class were lined up against each other and paired off into dancing partners. As it turned out, the arrangement was particularly fortuitous, for the difference in height between two partners in no case exceeded 4 inches.

Probably the most obvious way to avoid partners of unsuitable heights is to pair the shortest boy with the shortest girl, and so on down to the tallest boy with the tallest girl. Prove that, had this been done, the difference in the heights of any two partners would still not have exceeded 4 inches.

Solution by K. Seymour, Toronto, Ontario.   Obviously we shall have to use information gained from the lucky first way of arranging the partners to deduce the desired conclusion about the second, highly organized, scheme. Therefore, let us imagine that, *in both cases*, the boys are lined up in fixed order from shortest to tallest and that the distinction between the schemes lies in the way the girls are shuffled around in their two line-ups.

We may proceed nicely by the indirect approach. Suppose that some partners (b_i, g_i) in the orderly second arrangement differ in height by more than 4 inches; for definiteness, suppose that b_i is taller than g_i.

$$b_1 \ b_2 \ b_3 \ \cdots b_{i-1} \ b_i \ b_{i+1} \ \cdots b_{17}$$

$$g_1 \ g_2 \ g_3 \ \cdots g_{i-1} \ g_i \ g_{i+1} \ \cdots g_{17}$$

$$b_i - g_i > 4.$$

Now let the girls be moved back to their original partners. Clearly g_i can't remain with b_i since no partners in the first arrangement differ by more than 4 inches. In fact, if b_i is too tall for g_i, then he is also too tall for each of the girls $\{g_1, g_2, \ldots, g_{i-1}\}$, who are at best only as tall as g_i. That is to say, each of the i girls $\{g_1, g_2, \ldots, g_i\}$ must have been matched originally with one of the $i - 1$ shorter boys $\{b_1, b_2, \ldots, b_{i-1}\}$. Clearly this can only lead to trouble for some boy, and our conclusion follows.

(c)   The All-Union Olympiad

17.   The 8 integers $\{-1, -1, -1, -1, -1, -1, 2, 4\}$ have product $P = 8$ and sum $S = 0$. If the product P of n integers is equal to n itself and their sum $S = 0$, prove that n must be a multiple of 4, and that for every multiple of 4, $\{4, 8, 12, \ldots\}$, such a set of n integers does exist.

Solution by Bob Prielipp, University of Wisconsin-Oshkosh.   Suppose that $\{x_1, x_2, \ldots, x_n\}$ is a set of n integers having $P = n$ and $S = 0$. First of all we ask whether n could be odd. If n were odd, then P would be odd, and this would imply that every x_i would also have to be odd. But then S would be the sum of an odd number of odd integers and couldn't possibly be the even number 0. Hence n must be even, and can only escape being a multiple of 4 by being of the form $4k + 2$.

In the latter case, $n = P = 2(2k + 1) =$ twice an odd number, implying that one of the integers x_i is even and the remaining $n - 1$ of them are odd. Again, S is the sum of an odd number, $(n - 1)$, of odd integers (having an odd subtotal) and one even number, making it odd and not the prescribed value 0. Thus n must be a multiple of 4.

Every multiple of 4 is either 4 times an odd number $(4(2k - 1))$ or 4 times an even number $(4(2k))$. If $n = 4(2k - 1)$, it suffices to use the set

$$\{-1, -1, \ldots, -1, 1, 1, \ldots, 1, 2, -(4k - 2)\},$$

containing $2k - 1$ (-1)'s and $6k - 5$ $(+1)$'s; and when $n = 4(2k)$, we may use

$$\{-1, -1, \ldots, -1, 1, 1, \ldots, 1, 2, 4k\},$$

containing $6k$ (-1)'s and $2k - 2$ $(+1)$'s.

(d)   The All-Union Olympiad

22.   A math teacher wrote the quadratic trinomial $x^2 + 10x + 20$ on the blackboard. Then each student either increased by 1 or decreased by 1 either the coefficient of x or the absolute term. Finally the trinomial $x^2 + 20x + 10$ appeared. Did a quadratic trinomial with integer roots necessarily appear on the blackboard in the process?

Let the coefficient of x be a and the absolute term b; then the exercise at the blackboard began with $a = 10$, $b = 20$, and finished with $a = 20$, $b = 10$. That is to say, a started off at a value which is less than b and wound up being greater than b. Obviously, then, they must have "passed" each other at some point in the procedure. Since they change by increments of $+1$ or -1, a could not achieve a position of 10 units bigger than b without at some time being just 1 unit bigger than b. At this point the trinomial on the board would be of the form

$$x^2 + (k+1)x + k,$$

having integral roots of -1 and $-k$. Thus a trinomial with integral roots is unavoidable.

We should allow that a and b might pass each other at any integral value, not necessarily at an integer between 10 and 20. But the value of k is immaterial, for our claim is valid even for negative k: for example,

$$x^2 + (k+1)x + k = x^2 - 6x - 7,$$

still has integral roots -1 and $-k = 7$.

2. Unused Problems from International Olympiads (pp. 37–38)

(a)   Israel

1983 points are spaced evenly around a circle and each point is assigned either a $+1$ or a -1. Starting at any specified point, a partial sum is determined by adding up the $+1$'s and -1's around the circle as far as one may care to go. Thus for each starting point there are 1983 partial sums in the clockwise direction and another 1983 partial sums in the counterclockwise direction. Now, for some starting places in certain distributions of +1's and −1's, all 1983 partial sums in *both* directions are strictly positive—they start out on a positive note and never drop even to zero, no matter which way you choose to go around the circle. Prove that, if there are more than 1789 +1's on our circle, then there will be more than 1207 different starting places for which all 1983 partial sums in *each* direction will be strictly positive.

It's hard to suppress a chuckle when you read a question like this, and you can't help wondering where in the world they get the numbers 1789 and 1207 to go with the "number of the year" 1983. But if you do a little subtracting, you find that

$$1983 - 1789 = 194,$$

and that

$$1789 - 1207 = 582 = 3(194).$$

Rearranging the latter relation, we have

$$1789 - 3(194) = 1207.$$

Now, suppose there are n +1's and m −1's on the circle; then $n > 1789$ forcing $m < 194$, and it is a small step to use the result

$$1789 - 3(194) = 1207,$$

to get

$$n - 3m > 1207. \tag{A}$$

Clearly, if there were no −1's at all on the circle, the n +1's would provide n starting places that satisfy our requirements. The relation (A), then, strongly suggests that we should try to show that the addition of a −1 to the circle can never spoil more than 3 of these n "good" starting positions; if that can be done, the m −1's will not destroy more than $3m$

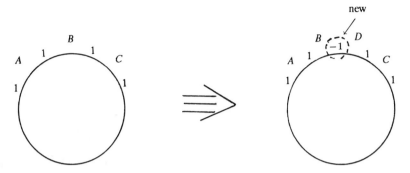

FIGURE 88
Introducing the −1 spoils A, B, and C (D is a new position, spoiled too, of course, due to the −1 itself).

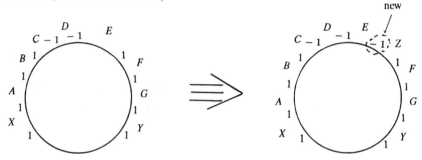

FIGURE 89
On the left, the 2 −1's already spoil A, B, C, D, E, F, and G; on the right, the new −1 extends the destructive effect of the string of −1's one place on each side, to X on the one side and Y on the other (besides its own new place Z)—if X and/or Y are already spoiled, the damage may be reduced.

of the original n good places, leaving at least $n - 3m$ of them unaffected, for a total exceeding 1207, as required.

    The problem is basically solved at this point, for it is now simply a matter of checking out the various cases to confirm our suspicions. Actually the worst case is obtained by putting a −1 in the midst of a string of +1's; in such a place it can spoil 4 starting places, but only 3 are among the original n good ones, the other one being introduced by the presence of the new element −1 itself. If a −1 is inserted next to an existing −1, it spoils at most 2 of the original n good places besides

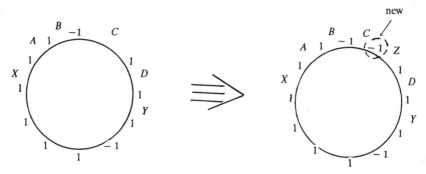

FIGURE 90
On the left, Y is already spoiled by the -1 nearest to it; on the right, X and (the brand new) Z are the only new places spoiled.

its own new place. I hope the accompanying figures are sufficient to convince you of these things.

We note that it is not necessary to spoil partial sums in both directions in order to ruin a starting position; either direction suffices.

(b)   U.S.A.

The sum of the face angles at all but one of the vertices of a given simple polyhedron P is $5160°$. What is the sum of the face angles at the omitted vertex?

Instead of going around the vertices, the sum S of the face angles of P can be determined by adding up the angles in all the faces. Therefore let the faces be numbered $1, 2, \ldots, F$ and let n_i denote the number of edges in the ith face. Since the sum of the interior angles in a k-gon is $(k-2)180°$, it follows that

$$S = \sum_{i=1}^{F}(n_i - 2)180 = 180\left[\sum_{i=1}^{F} n_i - 2F\right].$$

Now, in the sum $\sum_{i=1}^{F} n_i = n_1 + n_2 + \cdots + n_F$, each edge is accounted for twice, once in each of the two faces that it borders; hence, if E denotes the number of edges in P, this sum is just $2E$ and we have

$$S = 180(2E - 2F) = 360(E - F).$$

From Euler's famous formula $V - E + F = 2$, we get $E - F = V - 2$, and so

$$S = 360(V - 2).$$

If f denotes the required sum of the face angles at the omitted vertex, then clearly we also have

$$S = 5160 + f,$$

giving

$$S = 360(V - 2) = 5160 + f,$$
$$360V - 720 = 5160 + f$$
$$360V = 5880 + f.$$

But the sum of the face angles at any vertex must be less than 360° and we get the crucial result that

$$360V < 5880 + 360 = 6240, \qquad V < 17\frac{1}{3} \quad \text{and} \quad V \leq 17.$$

On the other hand, certainly

$$360V = 5880 + f$$

leads to

$$360V > 5880, \qquad V > 16\frac{1}{3}, \quad \text{and} \quad V \geq 17.$$

Consequently

$$V = 17,$$

and we have $360(17) = 5880 + f$, giving

$$f = 6120 - 5880$$
$$= 240 \text{ degrees.}$$

(c)   Canada (revised)

The sequence $\{2, 1, 2, 0, 0\}$ has the engaging property that it describes itself as far as the frequency of occurrence of its terms is concerned:

$$
\begin{array}{lccccc}
\text{Term} & 0 & 1 & 2 & 3 & 4 \\
\text{Frequency} & 2 & 1 & 2 & 0 & 0
\end{array};
$$

that is, there are 2 0's, one 1, 2 2's, and 0 3's and 4's. Other sequences having this remarkable characteristic are:

$$\{2, 0, 2, 0\}, \ \{1, 2, 1, 0\}, \quad \text{and} \quad \{6, 2, 1, 0, 0, 0, 1, 0, 0, 0\}.$$

In this rejected problem, Canada proposed that you find all possible sequences of this kind:

determine all finite sequences $\{n_0, n_1, n_2, \ldots, n_k\}$

such that the nonnegative integer i occurs exactly n_i times in the sequence.

An outstanding discussion of this subject is given by Michael D. McKay and Michael S. Waterman of the Los Alamos National Laboratory, New Mexico, in *The Mathematical Gazette* (March 1982, pages 1–4). Each step in the exposition is certainly straightforward and simple enough, but the road is fraught with the danger of becoming confused very easily unless one constantly reminds himself of the context in which the discussion is taking place. After carefully guiding the reader through a succession of little steps, McKay and Waterman arrive at the following general solution.

(i)   The only sequences of length less than 7 are the 3 examples given in the introduction.

(ii)   For each $k \geq 7$, there exists a *unique* sequence S_k of length k given by

$$S_k = \{k - 4, 2, 1, 0, 0, \ldots, 1, 0, 0, 0\};$$

only the first term varies with k; the second and third terms are always 2 and 1, and all the remaining terms are 0's except for a single 1 in the fourth last position (i.e., it always ends with $\cdots, 1, 0, 0, 0\}$). Thus

$$S_7 = \{3, 2, 1, 1, 0, 0, 0\},$$
$$S_8 = \{4, 2, 1, 0, 1, 0, 0, 0\},$$
$$S_9 = \{5, 2, 1, 0, 0, 1, 0, 0, 0\},$$

and so on.

You might also enjoy Tony Gardiner's fine pedagogical treatment of the subject (*The Mathematical Gazette*, March 1982, pp. 5–10), and a related paper by Lee Sallows and Victor L. Eijkhout (*The Mathematical Gazette*, March 1986, pp. 1–10).

3. The American Invitational Mathematics Exam, 1985 (p. 136)

13.   The numbers in the sequence 101, 104, 109, 116, ..., are generated from the formula $a_n = 100 + n^2$, $n = 1, 2, 3, \ldots$. For each n, let d_n be the greatest common divisor of the consecutive terms a_n and a_{n+1}. What is the *maximum* value taken by d_n as n runs through the positive integers?

To begin, we observe that

$$a_{n+1} = 100 + (n + 1)^2 = 100 + n^2 + 2n + 1 = a_n + 2n + 1.$$

Since d_n divides both a_{n+1} and a_n, d_n must also divide their difference, $2n + 1$. In this case, then

$$d_n \mid (2n + 1)^2 = 4n^2 + 4n + 1.$$

Now, because $d_n \mid a_n = 100 + n^2$, we also have $d_n \mid 4a_n = 400 + 4n^2$. Again, then, d_n must divide the difference

$$(400 + 4n^2) - (4n^2 + 4n + 1) = 400 - (4n + 1).$$

But clearly

$$d_n \mid 2(2n + 1) = 4n + 2,$$

and we have $d_n \mid [400 - (4n + 1)] + (4n + 2) = 401$, which is a prime number. Therefore d_n must actually be 1 or 401, a most surprising result.

It remains to determine whether any d_n actually has the value 401. Since $d_n \mid 2n + 1$, $n = 200$ suggests itself immediately and we find that

$$a_{200} = 40100 = 401(100), \quad a_{201} = 40501 = 401(101),$$

making $d_{200} = 401$, the required maximum value.

4. Dutch Olympiad, 1984 (p. 168)

1. (revised).   Suppose a circle C is drawn around a given unit circle A so as to touch it at the point P, as shown.

Let PQ be their common tangent at P and let a tangent to A in the perpendicular direction touch A at R, determine a chord MN in C, and meet PQ at S.

(i)   Prove that, however big the circle C might be, PR always *bisects* $\angle MPN$, and

(ii)   determine the radius r of C for which PR and PN will *trisect* $\angle MPS$.

To set the notation (see Figure 92), let A and C have centers O and K, and let angles MPR, RPN, NPS, and PRS, respectively be x, y, z, and t, as shown; let the perpendicular from K meet MN in T, and join N to K.

(i) Clearly the two tangents to A from S are equal, making PRS an isosceles right triangle; thus

$$y + z = t = 45°. \tag{1}$$

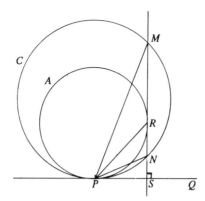

FIGURE 91

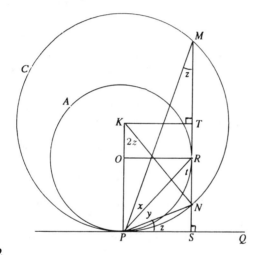

FIGURE 92

Now *the angle between a tangent and a chord of a circle is equal to the angle in the segment on the other side of the chord*, and so we have that

$$\angle SPN = \angle PMN \quad \text{in} \quad C, \quad \text{i.e.,} \quad \angle PMN = z.$$

Then, for $\triangle PMR$, the exterior angle $t = \angle PRN$ is equal to the sum of the two interior, opposite angles,

$$x + z = t \tag{2}$$

and from (1) and (2), we conclude that $x = y$, as required.

(ii) Suppose that PR and PN trisect $\angle MPS$, i.e., $x = y = z$. Since $y + z = 45°$ in $\triangle PRS$, then $y = z = 22\frac{1}{2}°$. In this case,

$$\angle PMN = z = 22\frac{1}{2}^{\circ},$$

and $\angle PKN$ at the center would be twice as big:

$$\angle PKN = 2 \cdot \angle PMN = 2z = 45°.$$

Now, clearly $PKTS$ is a rectangle and $PORS$ is a square. With $\angle PKN = 45°$, $\angle NKT = 45°$, and KTN is another isosceles right triangle. But $KT = PS = OP = 1$, and so $KT = TN = 1$, and the radius r of C is given by $r = KN = \sqrt{2}$ (by Pythagoras). That is to say, if PR and PN trisect $\angle MPS$, then $r = \sqrt{2}$. However, the question that is asked concerns the converse result. We still need to show that if $r = \sqrt{2}$, it follows that $x = y = z$.

Fortunately, this is easily checked: if $KN = \sqrt{2}$, then in right triangle KTN, the pythagorean theorem gives

$$TN^2 = KN^2 - KT^2 = 2 - 1 = 1, \quad \text{giving} \quad TN = 1,$$

and we have $\triangle KTN$ is isosceles, implying that $\angle NKT = 45°$; then $2z = \angle NKP = 45°$, making $\angle PMN = z = 22\frac{1}{2}°$; since $y + z = 45°$, then $y = 22\frac{1}{2}°$, and since $x = y$, we have $x = y = z = 22\frac{1}{2}°$.

Thus when C has radius $\sqrt{2}$, PR and PN will actually be the trisectors of $\angle MPS$.

5. Bulgarian Winter Competition, 1985, Grade 9, (p. 270)

2. R and B are two disjoint nonempty finite sets of points in the plane; each point of R is colored red and each point of B is colored blue. If every *segment* which joins two points of the same color contains a point of the other color, prove that all the points of the two sets must lie on a single straight line.

We proceed indirectly. Suppose that some 3 points of $A \cup B$ determine a nondegenerate triangle. Since A and B are finite sets, the number of nondegenerate triangles is finite, and hence there must be a nondegenerate triangle XYZ of *minimum area*.

Now, there are only two easy cases.

(i)   *All 3 of X, Y, Z are the same color* (say red).

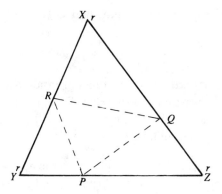

FIGURE 93

In this case, each side of $\triangle XYZ$ contains a blue point, and a smaller nondegenerate blue triangle PQR is determined, contradicting the minimum character of $\triangle XYZ$.

(ii)   *Only 2 of X, Y, Z are the same color* (say Y and Z are red).

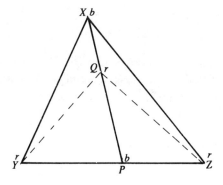

FIGURE 94

On the side YZ there must be a blue point P somewhere between the red vertices, and then on the "blue" segment XP there will be a red point Q, giving the contradiction of a smaller nondegenerate triangle QYZ. Thus $A \cup B$ cannot contain a nondegenerate triangle and must be confined to a single straight line.

6. Unused Problems—the International Olympiad, 1985 (pp. 304–307)

(a) Israel

12. 1985 points are chosen at random inside a unit cube. Show that it is always possible to select 32 of the chosen points so that, in whatever order they may be taken, the perimeter of the 32-gon they determine is less than $8\sqrt{3}$.

Suspecting that the pigeonhole principle was at the heart of things, I began this problem by dividing 1985 by 31 (not 32) and got a result that was just a little over 64 (actually $64.0322\ldots$). Turned around, this implies that 1985 divided by 64 is just a little bigger than 31 (in fact $31.0156\ldots$). Therefore, if the unit cube were to be partitioned into 64 sections and 1985 points chosen, some section would have to contain more than 31 of the points, i.e., at least 32 of them. From here the solution was pretty clear.

The obvious way to divide the cube into 64 parts is by quartering each edge and slicing it into 64 little $1/4 \times 1/4 \times 1/4$ cubes. Two points that belong to the same little cube cannot be farther apart than

$$d = \sqrt{(1/4)^2 + (1/4)^2 + (1/4)^2} = \sqrt{\frac{3}{16}} = \frac{1}{4}\sqrt{3},$$

and only a pair of diametrically opposite vertices can actually be this far apart. Thus the perimeter of any 32-gon contained in a little cube could not even be as great as $32(\frac{1}{4}\sqrt{3}) = 8\sqrt{3}$, since only 8 of its vertices could be vertices of the little cube itself.

(b) Czechoslovakia

5. (The rejected proposal was the 3-dimensional version of the following 2-dimensional problem.)   Consider the set T of

all lattice points (x, y) in the plane (i.e., x and y both integers).
The *neighbors* of (x, y) are the 4 lattice points that lie at a unit
distance from (x, y), namely $(x \pm 1, y)$ and $(x, y \pm 1)$. A lattice
point and its 4 neighbors, then, comprise a little clique C of 5
points. Prove that there exists a subset S of the lattice points
which has exactly one point in common with each clique in the
plane (that is, to each clique C, S contributes either the primary point (x, y) itself and none of the neighbors, or else precisely one of the neighbors and not the primary point (x, y)).

It strikes me as very interesting that such a subset as S should exist at
all. With a little trial and error, one soon arrives at a solution

$$S = \{(x, 2x + 5k) \mid x, k \text{ integers}\}.$$

An illustration is probably enough to convince one that S *is* a solution,
but the proof, which is not very exciting, is left to the reader who is
especially interested.

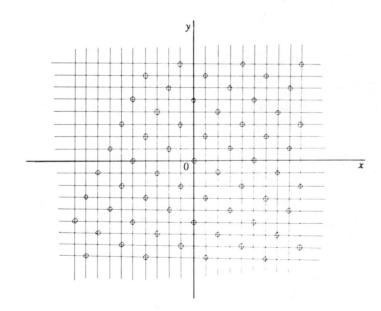

FIGURE 95

DIOPHANTINE RECIPROCALS

How many solutions in positive integers (x, y) are there to the equation

$$\frac{1}{x} + \frac{1}{y} = \frac{1}{n}, \quad n \text{ a positive integer?}$$

Solution

Once an acceptable value is assigned to either x or y, there is no choice for the other. In terms of x, we have

$$\frac{1}{y} = \frac{1}{n} - \frac{1}{x} = \frac{x - n}{nx},$$

giving

$$y = \frac{nx}{x - n},$$

and the solution

$$(x, y) = (x, \frac{nx}{x - n}).$$

The problem, then, is to count the positive integers x that make $x - n$ a positive divisor of nx, where n is a given positive integer.

Clearly this requires $x - n \geq 1$, i.e., $x \geq n+1$. But with x figuring in the numerator and denominator of y, both these numbers vary with x, and it is not evident how to determine or count the acceptable integers x.

It would appear to be a step in the right direction to get rid of x from one of these quantities. Proceeding with a maneuver as common as long division, we obtain

$$y = \frac{xn}{x - n} = \frac{[(x - n)n + n^2]}{x - n} = n + \frac{n^2}{x - n}.$$

Thus, in order to make y an integer, $x - n$ must be a divisor of the constant n^2; we have already noted that $x - n$ must be positive. In fact, if the positive integer k divides n^2, then, setting $x - n = k$, we obtain a solution

$$(x, y) = (n + k, n + \frac{n^2}{k})$$

to our equation. Hence the number of solutions N is simply the number of positive divisors k of n^2.

If the prime decomposition of n is

$$n = p_1^{a_1} p_2^{a_2} \cdots p_k^{a_k},$$

then

$$n^2 = p_1^{2a_1} p_2^{2a_2} \cdots p_k^{2a_k},$$

and the number of solutions N is given by

$$N = (2a_1 + 1)(2a_2 + 1) \cdots (2a_k + 1).$$

ANOTHER SERIES OF RECIPROCALS

Let

$$S_n = 1 + \frac{1}{2} + \frac{1}{3} + \cdots + \frac{1}{n},$$

the nth partial sum of the famous harmonic series H. Because H diverges, the reciprocals $1/S_n$ approach 0 as n increases without limit. Consequently the convergence of the series S, determined by the sum of these reciprocals,

$$S = \sum_{n=1}^{\infty} \frac{1}{S_n},$$

is an open question. It is known that the harmonic series diverges very slowly, which tends to keep S_n small and $1/S_n$ large, and thus favors the divergence of S. Determine whether S diverges or not.

Solution

Clearly, for $n > 1$,

$$S_n = 1 + \frac{1}{2} + \cdots + \frac{1}{n} < 1 + 1 + \cdots + 1 = n.$$

Thus $1/S_n > 1/n$, and so S is even bigger, term for term, than H, implying that S is indeed divergent after all.

AN ILLEGIBLE
MULTIPLE-CHOICE PROBLEM

In a certain multiple-choice test, one of the questions was illegible, but the choice of answers, given below, was clearly printed. What is the right answer?

(a) All of the below.
(b) None of the below.
(c) All of the above.
(d) One of the above.
(e) None of the above.
(f) None of the above.

Solution

There is no big point to make in this problem. It is just one of those intriguing little problems that is somewhat out of the ordinary and you might wonder at first whether it can be done. When you come to try it, you find there is no more to it than a straightforward process of elimination, but still there is some satisfaction simply in seeing that the question is feasible.

First of all, the validity of (a) implies that all the answers (b) to (f) are true; but if (b) is valid, then none of the rest is, and we have a contradiction. Hence (a) must be false.

Because (c) claims (a) to be true, we get that (c) must be false, too.

In this case, the validity of (b) would mean that (b) is the only valid answer among the first three, thus confirming (d), which, in turn, contradicts (b) itself. Hence (b) must be false, making all three of (a), (b), and (c) false. And, because this fact denies (d), it must be that (d) also is false.

Now (f) cannot be true, for if it were, then all of (a) to (e) would be false, confirming (e) (a contradiction). Thus the right answer has to be (e), and this is confirmed by the falsity of (a) to (d).

ON THE PARTITION FUNCTION $p(n)$

Let $p(n)$ be the number of unordered partitions of the positive integer n, that is, the number of ways of writing n as an unordered sum of one or more positive integers. For example, $p(4) = 5$ because of the 5 sums

$$4, \quad 3+1, \quad 2+2, \quad 2+1+1, \quad 1+1+1+1.$$

The function $p(n)$ increases *very* rapidly with n; for example,

$$p(10) = 42, \quad p(20) = 627, \quad p(50) = 204226,$$

and

$$p(200) \text{ is almost 4 } \textit{trillion}.$$

There is no known formula which gives $p(n)$ in terms of n.

Now the special characteristic of a "convex" function f is that its value at the midpoint of every interval $[a, b]$ in its domain does not exceed the average of its values at the ends of the interval.

Prove that the partition function $p(n)$, with the obvious minor allowances for its discrete character (n must be a positive integer), shares

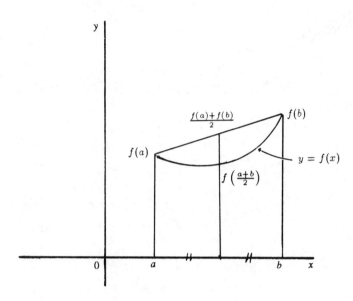

FIGURE 96

to a limited extent in this basic property of convexity:

$$p(n) \leq \frac{p(n+1) + p(n-1)}{2}.$$

Solution

This condition may be arranged to give

$$p(n+1) + p(n-1) \geq 2p(n),$$

and

$$p(n+1) - p(n) \geq p(n) - p(n-1).$$

Consequently, let's have a look at the partitions of a pair of consecutive integers n and $n+1$.

If 1 is added as an extra part at the end of a partition of n, a partition of $n+1$ is obtained, in particular a partition that contains a 1.

Conversely, if a 1 is deleted from a partition of $n + 1$ which contains a 1, a partition of n results. Thus there is a 1-1 correspondence between the entire set of partitions of n and the subset of partitions of $n + 1$ that contain a 1. Hence $p(n + 1) - p(n)$ is the number of partitions of $n + 1$ in the subset X of partitions which do *not* contain a 1.

Similarly, $p(n) - p(n - 1)$ is the number of partitions of n in the subset Y of partitions which do not contain a 1. If we can show that $|X| \geq |Y|$, the desired conclusion follows. This simply requires we show that, to each partition in Y, there corresponds a *distinct* partition in X.

$n = 4$	$n + 1 = 5$
4	4+1
3+1	3+1+1
2+2	2+2+1
2+1+1	2+1+1+1
$1 + 1 + 1 + 1$	$1 + 1 + 1 + 1 + 1$
	5
	3+2

$$Y = \{4, 2 + 2\} ; \quad X = \{5, 3 + 2\}.$$

There is a small pitfall here that must be avoided. If we add 1 to a partition in Y, we get a partition of $n + 1$ alright, but it won't belong to X if the 1 is added as an *extra part* (the partitions in X don't contain 1). Accordingly, let us include the 1 by adding it to one of the parts already in the partition. Unfortunately, unless this is done with care, it won't always yield *distinct* partitions of $n + 1$. For example, for $n = 13$, both $(4, 3, 3, 3)$ and $(4, 4, 3, 2)$ would belong to Y; increasing the 3 in the first and the 2 in the latter both give $(4, 4, 3, 3)$ in X. However, if we always increased the *greatest* part in a partition in Y, then distinct partitions in X are guaranteed; clearly, by this method, identical partitions in X could only be generated by identical antecedents in Y. Hence $|X| \geq |Y|$, and our solution is complete.

GLEANINGS FROM MURRAY KLAMKIN'S OLYMPIAD CORNERS—1986

1. Dutch Olympiad, First Round, 1985 (p. 3)

The terms a_n of a sequence of positive integers satisfy

$$a_{n+3} = a_{n+2}(a_{n+1} + a_n) \quad \text{for} \quad n = 1, 2, 3, \ldots.$$

If $a_6 = 144$, what is a_7?

Solution by Andy Liu, University of Alberta. Obviously the first three terms of the sequence need to be prescribed in order to seed the recursion; suppose

$$a_1 = x, \quad a_2 = y, \quad a_3 = z.$$

Then

$$a_4 = z(y + x),$$
$$a_5 = z(y + x)(z + y), \quad \text{and}$$
$$a_6 = z(y + x)(z + y)[z(y + x) + z]$$
$$= z^2(y + x)(z + y)(y + x + 1).$$

Admittedly this does not look encouraging.

However, if you happen to pick up on the fact that the factors $(y + x)$ and $(y + x + 1)$ are *consecutive* integers, the solution follows readily. Listing the divisors of $144 = 2^4 \cdot 3^2$, we have

$$1, \ 2, \ 3, \ 4, \ 6, \ 8, \ 9, \ 12, \ 16, \ 18, \ 24, \ 36, \ 72, \ 144,$$

containing only the four pairs of consecutive integers $(1, 2)$, $(2, 3)$, $(3, 4)$, and $(8, 9)$. Now the smallest x, y, and z can be is 1, making $(y + x, y + x + 1)$ at least the pair $(2, 3)$. It remains only to check the feasibility of the three possibilities.

For $(2, 3)$, it must be that $x = y = 1$, and then we have

$$144 = z^2(2)(z + 1)(3),$$
$$24 = z^2(z + 1),$$

which has no integral solution for z.

For $(3, 4)$, we have

$$144 = z^2(3)(z + y)(4)$$
$$12 = z^2(z + y),$$

which forces $z = 2$ and $y = 1$ ($z = 1$, $y = 11$ is inadmissible since $y + x = 3$); this yields $x = 2$, and we get

$$a_7 = a_6(a_5 + a_4) = 144[2(3)(3) + 2(3)] = 144(24) = 3456.$$

Finally, the pair $(8, 9)$ yields

$$144 = z^2(8)(z + y)(9)$$
$$2 = z^2(z + y),$$

implying $z = 1$, $y = 1$, and $x = 7$ (since $x + y = 8$), giving

$$a_7 = 144(16 + 8) = 3456.$$

Thus a_7 can only be 3456.

2. Four Problems from the American Invitational Mathematics Exam, 1986 (pp. 67–68)

(a)   What is the *greatest* positive integer n which makes $n^3 + 100$ divisible by $n + 10$?

If that 100 were only 1000, we could factor the expression as the sum of two cubes. The thing to do, then, is to make it 1000 by adding and subtracting 900:

$$n^3 + 100 = (n^3 + 1000) - 900.$$

Thus, for $n + 10$ to divide $n^3 + 100$, it must also divide 900. Clearly the greatest integer to do that is 900, itself, and we have the greatest n must be 890.

(b)   A student practiced his arithmetic by adding up the page numbers of a book and he got a total of 1986. Unfortunately he was interrupted during the exercise and had inadvertently added in one of the pages twice. What page was it?

If the pages were numbered $1, 2, \ldots, n$, then the correct sum would have been $n(n + 1)/2$. Since his mistake increased his total, we have that

$$\frac{n(n + 1)}{2} < 1986, \quad \text{and} \quad n(n + 1) < 3972.$$

Now $n(n+1)$ is reasonably approximated by n^2, and so we would expect n to be in the vicinity of $\sqrt{3972} = 63.02\ldots$. Since $63(64) = 4032$, n can't be as big as 63, but $n = 62$ seems to be feasible, for $62(63) = 3906$. Because the error E is a page number that is $\leq n$, then $E \leq 62$, which implies that $n = 61$ is too small as follows: $\frac{1}{2}(61)(62) = 1891$, which even with the maximum error added in, gives only $1891 + 62 = 1953$, falling short of the grand total 1986. Thus n must be 62, and the correct total is $62(63)/2 = 1953$. Hence it was page 33 that was counted twice.

(c)   The increasing sequence 1, 3, 4, 9, 10, 12, 13, ..., consists of those positive integers which are powers of 3 or sums of distinct powers of 3. What is the 100th term of this sequence?

If the terms of the sequence are written in base 3 (where the digits are only 0, 1, or 2), we see that they comprise the positive integers which do *not contain the digit 2* (the powers of 3 must be *distinct*):

$$1, \ 10, \ 11, \ 100, \ 101, \ 110, \ 111, \ldots .$$

Of course, in the *binary* scale these numerals are simply the consecutive integers $1, 2, 3, \ldots$. The 100th one, then, is 1100100, the binary number for 100, and in its ternary interpretation, it represents the integer

$$3^6 + 3^5 + 3^2 = 729 + 243 + 9 = 981.$$

(d)   If S is the sum of the common logarithms of all the proper divisors of a million, what is the integer nearest S?

The prime decomposition of a million is simply $10^6 = 2^6 \cdot 5^6$. Clearly, then, any integer d of the form $2^a \cdot 3^b$, where $0 \le a, \, b \le 6$, is a divisor of a million. Since there are 7 choices for each of a and b, the *number* of divisors of a million is 49.

Now the divisors of any integer go together into complementary pairs $(d, n/d)$; and since $\log d + \log \frac{n}{d} = \log[d \cdot \frac{n}{d}] = \log n$, the sum of the common logarithms of a pair of complementary divisors of a million is just 6. Our 49 divisors go together into 24 such pairs with the self-complementary square root of a million, namely 1000, left over. Thus the sum of the common logarithms of *all* the divisors of a million is

$$24(6) + 3 = 147.$$

Deducting 6 for the *improper* divisor 10^6, itself, the required sum S for the *proper* divisors is just 141.

3. Spanish Olympiad, First Round 1985 (p. 97)

If n is a positive integer, prove that $(n + 1)(n + 2) \cdots (2n)$ is divisible by 2^n.

Although it would appear to be a move in the wrong direction, the thing to do here is to multiply by $n!/n!$; this gives

$$(n+1)(n+2)\cdots(2n)\cdot\frac{n!}{n!}$$

$$=\frac{1\cdot2\cdot3\cdots(2n-1)(2n)}{n!}$$

$$=\frac{[1\cdot3\cdot5\cdots(2n-1)][2\cdot4\cdot6\cdots(2n)]}{n!}$$

$$-\frac{[1\cdot3\cdot5\cdots(2n-1)][2^n(1\cdot2\cdot3\cdots n)]}{1\cdot2\cdot3\cdots n}$$

$$=2^n[1\cdot3\cdot5\cdots(2n-1)].$$

The conclusion follows.

4. A Weierstrass Product Inequality, (p. 273)

If $0\le a,b,c,d\le 1$, prove that

$$(1-a)(1-b)(1-c)(1-d)+a+b+c+d\ge 1.$$

Solution by Murray Klamkin.   Let the function on the left side of the inequality be called $f(a,b,c,d)$. Obviously it is a function of four variables. Suppose that all but one of the variables, say b,c, and d, are set at some fixed values and a is allowed to vary over its range from 0 to 1. With b,c, and d fixed, f is just a function of a:

$$f(a)=(1-a)K+a+M,$$

where K and M are constants determined by the fixed values of b,c, and d. In fact, $f(a)$ is a *linear* function of a, with a *linear* graph, and it is clear from the graph that it attains a minimum value at an endpoint of its range, i.e., at $a=0$ or $a=1$; and, within useful limits, we can deduce which it is.

Rearranging things, we have

$$f(a)=(1-K)a+K+M,$$

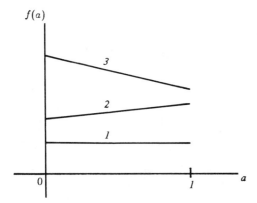

FIGURE 97

where $0 \leq K = (1 - b)(1 - c)(1 - d) \leq 1$, since $0 \leq b, c, d \leq 1$. Thus the slope $1 - K$ of the graph cannot be negative, eliminating case 3 that is shown in the illustration. It is immaterial whether the slope is zero or positive, for in either case the minimum value of $f(a)$ occurs at $a = 0$.

But the inequality does not favor the variable a; the same result must also hold for b, c, and d, and we conclude that the minimum value of $f(a, b, c, d)$ is given by $f(0, 0, 0, 0)$ (perhaps not *uniquely*, but that doesn't matter).

Accordingly,

$$f(a, b, c, d) \geq f(0, 0, 0, 0)$$
$$= 1 \cdot 1 \cdot 1 \cdot 1 + 0 + 0 + 0 + 0$$
$$= 1,$$

as required.

Of course one of the beauties of this approach is that the number of variables is inconsequential and therefore, without further ado, all other cases follow as immediate corollaries.

PROOF BY INTERPRETATION

It is always a special pleasure to establish a relation by interpreting it as a simple characteristic of a geometrical, combinatorial, or other kind of model. For example, the identity

$$\sum_{k=0}^{r} \binom{m}{k}\binom{w}{r-k} = \binom{m}{0}\binom{w}{r} + \binom{m}{1}\binom{w}{r-1} + \cdots + \binom{m}{r}\binom{w}{0}$$
$$= \binom{m+w}{r},$$

is proved most elegantly by noting that each side merely counts the number of committees consisting of r people that can be formed from m men and w women: the right side gives the result all at once, while on the left the various terms $\binom{m}{k}\binom{w}{r-k}$ respectively count the times that the committee is composed of k men and $r-k$ women.

Determine an experiment in probability to justify the inequality

$$(1 - p^m)^n + (1 - q^n)^m > 1$$

for all positive integers m and n greater than 1 and all positive real numbers p and q such that $p + q \leq 1$.

246

Solution

In any triangle ABC, let two fans of straight lines across the triangle issue from vertices B and C, n lines from B and m from C (the case $n = 3$, $m = 2$ is shown). At each of the mn points of intersection determined by the fans draw a little circle centered at the point. In each circle, then, there is a "cross hair" produced by the fans, one hair belonging to a line from each fan.

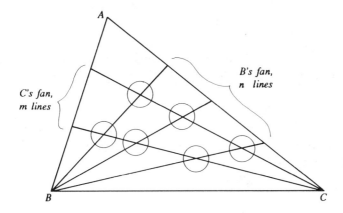

C's fan,
m lines

B's fan,
n lines

A

B   C

FIGURE 98

Now let the following experiment be performed. In each circle, delete one of the hairs, thus puncturing a line in one of the fans, and leave the other hair intact, thus preserving, at least in that region, the unbroken character of the line from the other fan. In some experiments none of the $m+n$ lines in the fans get through without suffering a break somewhere; in others, many lines do so.

Now suppose the experiment is performed so that, for each circle, the probability is p that the deleted hair breaks the line in C's fan and $q = 1 - p$ that it breaks the line in B's fan; thus p is the probability that B's line remains unbroken at this point and q the probability that C's line does so. Since each line in the fan from B passes through m little circles, the probability that it is not broken somewhere along the line is p^m and so the probability that it *does* get punctured somewhere is

$1 - p^m$. Hence the probability that all n of the lines in B's fan are broken somewhere is $(1 - p^m)^n$. Finally, the probability that this is *not* the case, that is, that at least one of the lines in B's fan survives completely intact is

$$1 - (1 - p^m)^n.$$

Similarly, the probability that at least one of C's lines emerges unscathed is

$$1 - (1 - q^n)^m.$$

Now the key point here is that a completely unbroken line from either fan can exist only at the expense of breaking *every* line in the other fan.

Let's also use just B and C for the events that at least one of their lines remains completely unbroken. Then B and C are *mutually exclusive* events. Because it concerns just *disjoint* subsets of outcomes, we have the basic result

$$pr(B) + pr(C) \le 1.$$

As it turns out, however, the events B and C don't encompass all the possibilities, for there is always an experiment of nonzero probability that breaks all the lines of both fans. As we shall soon see, this is a consequence of the requirement that m and n both exceed 1. Accepting this momentarily, we have

$$pr(B) + pr(C) < 1,$$

giving

$$[1 - (1 - p^m)^n] + [1 - (1 - q^n)^m] < 1,$$

showing that

$$1 < (1 - p^m)^n + (1 - q^n)^m$$

holds for $q = 1 - p$, i.e., for $p + q = 1$. Since a decrease in p or q merely strengthens this relation, it holds also for all p, q such that $p + q \le 1$.

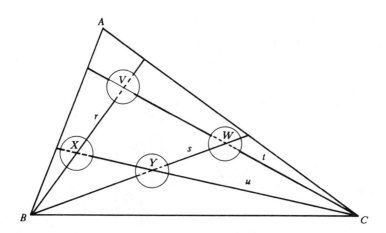

FIGURE **99**

We conclude with the description of an outcome which does not belong to either of the events B or C, showing that they are indeed not all- inclusive.

Let r and s be any two lines from B's fan (recall $n > 1$) and let t and u be any two lines from C's fan ($m > 1$), meeting at the four points V, W, X, Y as shown.

Now, as the deletions are performed in the circles along the line t, let all lines from B be broken except r and s; the circles at V and W will be attended to later. Similarly, proceed along r, destroying all the lines from C except t and u, leaving V and X. Thus all the lines except r, s, t and u are already broken, and these remaining four lines may be broken as indicated by the dotted hairs in the figure.

ON $\sigma(n)$ AND $\tau(n)$

Let $\sigma(n)$ denote the *sum* of the positive divisors of the positive integer n and $\tau(n)$ the *number* of such divisors.

Prove the engaging relation that

$$\tau(n) \quad \text{is a prime number whenever} \quad \sigma(n) \quad \text{is.}$$

Solution

If $n > 1$, then $\sigma(n) > 1$; thus, for $a, b > 1$, the number $\sigma(a) \cdot \sigma(b)$ is the product of two integers each > 1 and is therefore a composite number. Now it is well known that the function $\sigma(n)$ is multiplicative, that is, for relatively prime a and b,

$$\sigma(ab) = \sigma(a) \cdot \sigma(b).$$

Consequently, if n is divisible by two or more *different* primes, then n can be factored into some pair of relatively prime integers a and b, each > 1, and it follows that $\sigma(n) = \sigma(ab)$ must be a composite number. Consequently, $\sigma(n)$ can be a *prime* number only when n has just one prime divisor, making $n = p^k$, a power of some prime p.

It follows that

$$\sigma(n) = 1 + p + p^2 + \cdots + p^k = \frac{p^{k+1} - 1}{p - 1},$$

and it is far from clear how we can make good use of the knowledge that this number is a prime.

Perhaps this is a job for the contrapositive approach. Accordingly, let's see if it is any easier to get something out of the supposition

"$\tau(n)$ is *not* a prime number."

In this case, $\tau(n) = ab$, for some integers $a, b > 1$. Now, because $n = p^k$, a direct count of its divisors $(1, p, p^2, \ldots, p^k)$ gives

$$\tau(n) = k + 1,$$

and we have

$$k + 1 = ab.$$

Then we have

$$\sigma(n) = \frac{p^{k+1} - 1}{p - 1} = \frac{p^{ab} - 1}{p - 1} = \left[\frac{(p^a)^b - 1}{p^a - 1} \right] \left[\frac{p^a - 1}{p - 1} \right],$$

which is the product of two integers > 1, since $a, b > 1$, (each denominator is clearly a factor of its numerator), contradicting the primality of $\sigma(n)$. Thus if $\sigma(n)$ is *known* to be a prime number, it must be that $\tau(n)$ is one, too.

A SURPRISING RESULT
ABOUT TILING A RECTANGLE

The lead article in the 1987 August-September issue of the *American Mathematical Monthly* is entitled "Fourteen Proofs of a Result About Tiling a Rectangle" (by Stan Wagon, Smith College, Northampton, Massachusetts). At the level of sophistication of the readers of the *Monthly*, many of the proofs are only a few lines long. In this morsel I would like to expand on two of the most beautiful of these proofs.

1. The Result

Suppose a rectangle R is covered exactly with smaller rectangles, without gaps or overlapping. These covering tiles may be of assorted sizes, with or without repetitions, and it is immaterial how many of them are used. It is pretty obvious that a covering of R could never be completed if any tile were tilted so that its sides were not parallel to the sides of R. Thus if R runs horizontally and vertically, each tile must do the same, and there are only two ways of orienting a given tile—either across or up-and-down (in fact, there is only one way for a square tile).

We begin with a rectangle R that is tiled with some collection of little rectangles. In general, the dimensions of a tile can be any lengths whatever. The kind of tiling we wish to consider, however, is special in that every tile in it has a side of *integer* length; the other side may or may not also be an integer. And that's all we are told about the tiling.

For any given tile, we don't know whether it's the horizontal dimension that is an integer or the vertical (or possibly both); all we know is that at least one of them is an integer. Generally there are many noninteger sides across R among those which add up to its total width and many which contribute to its total height. In view of this, it may be somewhat surprising to learn that

Theorem. *If a rectangle $R = OABC$ can be tiled with rectangles each of which has an integer side, then R itself must have an integer side.*

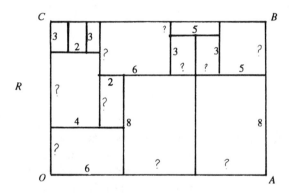

FIGURE **100**

2. A Proof From Graph Theory

Our first step is to assign cartesian coordinates to the figure, setting the origin at the corner O and the x-axis along the side OA, so that R fits snugly against the axes in the first quadrant.

Now some tile M sits in the corner at O and has two of its sides OD and OL run along the coordinate axes. Since at least one of these sides of M is an integer, then at least one of its corners D and L has integer coordinates—either D is $(q, 0)$ for some integer q and/or L is $(0, r)$ for some integer r; in the example, D is (6,0). Similarly, the tile N which is adjacent to M at such a corner has at least one of its sides DT or

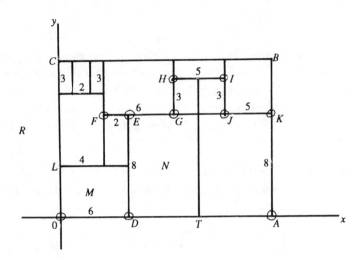

FIGURE 101

DE an integer, making the two coordinates of at least one of T and E both integers; (E is (6,8)). In the example, this identification of lattice points (i.e., both coordinates integers) can be continued through the chain $O(0,0)$, $D(6,0)$, $E(6,8)$, $F(4,8)$, $G(10,8)$, $H(10,11)$, $I(15,11)$, $J(15,8)$, $K(20,8)$, right to $A(20,0)$, showing that the side OA of R is indeed an integer. In general, if the collection S of lattice points which are to be found among the corners of the tiles should contain any of the vertices A, B, or C of R, then clearly R must have an integer side (S always contains the origin O). We shall see that this is indeed always the case by a beautiful application of the most elementary graph theory.

Let a graph Z be constructed to have a vertex for each tile in the covering of R and also a vertex for each lattice point in the collection S of lattice points that occur among the corners of the tiles. Let the vertices which represent the tiles $(M, N, U, V, \ldots)$ be lined up in one row P and those representing the lattice points $(O, D, E, F, \ldots)$ in another row Q.

Now the edges are put in so that each joins a vertex in P to a vertex in Q (making Z a bipartite graph). The rule for putting in the edges is as follows:

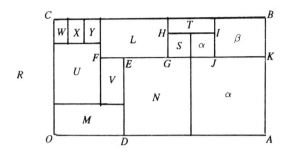

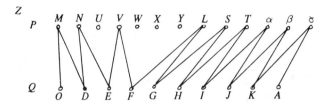

a vertex v in P is joined to a vertex u in Q if and only if the lattice point that is represented by u is a corner of the tile represented by v; thus, one simply goes from tile to tile, noting which of its 4 corners belongs to the collection S of lattice points.

Suppose, in the example, that the set S of lattice points contains only the corners we have mentioned: $S = \{O, D, E, F, \ldots, A\}$. In the graph Z, then, M is joined to O and to D, N to D and E, U is left unjoined, as are W, X, and Y, L is joined to F and G (not to E or H, though, since they are not corners of L), and so on.

Now then, let us count the edges in this graph Z. Since each edge has one end in row P and the other in row Q, the number of edges is given by both the total number of endpoints occurring at the vertices in P and the total number of endpoints occurring at the vertices in Q. That is to say, these total numbers of endpoints for P and Q are equal. This is the fundamental relation that yields the desired result.

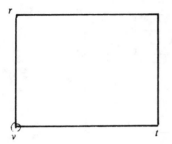 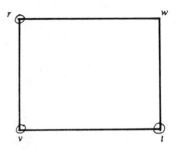

FIGURE 103

Before using this equality, however, we need to observe that the total in question will always be an *even* number. The crucial point is that no tile can have exactly 1 or exactly 3 lattice points among its 4 corners.

If one corner v is a lattice point, then, because at least one side is an integer, at least one of the adjacent corners r and t will inherit two integer coordinates; and 3 lattice corners v, r, and t similarly force a fourth lattice corner. Consequently, as we count up the endpoints at the vertices in P, each contribution to the sum is either a 0, 2, or 4, and an even total must result. Therefore, when we get to Q, we know already that the total we shall reach will be an *even* number.

Now the number of edges at a vertex v in Q is the number of times the lattice point represented by v is a corner of a tile. Specifically, then, we can see that the vertex representing the origin will always have just one edge at it and that, in general, any lattice point which is not a corner of the frame R itself (that is, neither O, A, B, or C) must have either 2 or 4 edges at it. There is simply no way for a point, which is not a corner of R, to be a vertex of exactly 1 or exactly 3 abutting rectangles (0 or 1 abutments are impossible, and 3 abutments force a fourth).

Among the contributions to the even total number of endpoints in Q, we have a 1 from the vertex representing the origin and a 2 or a 4 from each vertex that does not represent a corner of R:

$$1 + 2 + 2 + 4 + 2 + \cdots = \text{ an } even \text{ total.}$$

Clearly there must be another odd contribution by some vertex in Q in order to yield a grand total that is even. Since the corners A, B, C

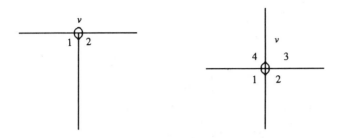

FIGURE **104**

of R are the only corners we haven't considered (they are also corners of little tiles), they are the only possible source of this necessary odd contribution, implying that at least one of A, B, C must actually be represented in Q. Thus at least one of A, B, C must be a lattice point and our conclusion follows.

Comment.   We should note that, from our scanty knowledge of a tiling, we would not actually be able to construct the graph Z. We know, for example, that one of the corners D, L of M is a lattice point, but we don't know which one; in fact, we never know specifically which edges to put in. But this doesn't mean that the graph Z doesn't exist. It's just that we are unable to produce it. Fortunately we do not need to have Z at hand in order to *deduce* a couple of the basic properties that it must have. The fact that, corresponding to every tiling, there *exists* a graph Z is all we need in order to make our theoretical observations.

7. A Checkerboard Proof

Due to Richard Rochberg, Washington University, and Sherman Stein, University of California, Davis.   Instead of R occurring in the first quadrant of a coordinate plane, this time suppose that R is put in the lower left corner of a "large" checkerboard whose squares have sides of length $\frac{1}{2}$.

Now it is not difficult to convince oneself that a segment of *integer* length, placed anywhere on the board in a position parallel to an edge of the board, must have a total of half its length running over black regions and half through white regions. Being of integral length, it would be ex-

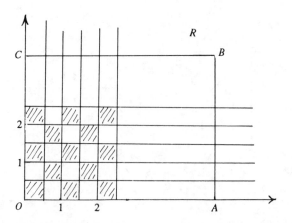

FIGURE 105

actly long enough to span an *even* number of complete rows or columns of the board (they're only $\frac{1}{2}$ a unit wide); sliding it perpendicularly to itself doesn't alter things until it enters the next row or column, in which case the colors reverse but still maintain a 50-50 split; and sliding it in its own direction causes it to pick up at one end an equal amount of the same color that it loses at the other end, again preserving that equality of the totals of black and white parts. But each rectangular tile is just the region traced out by sliding its integer side across the tile to the opposite side. Because such a sweepline covers the same amount of black and white in all its positions, the tile itself must cover the same amount of black area as white. And since R can be exactly covered by such tiles, then R, too, must cover the same amount of black as white.

Now we proceed indirectly. Suppose neither side of R is an integer. Then the edges AB and CB, respectively, slice through the interiors of a column and row of squares on the board. If a is the greatest integer less than OA and c the greatest integer less than OC, R contains a complete $a \times c$ checkerboard I in its bottom left corner. If the edges of this subboard are extended, R is divided into 4 pieces I, II, III, IV, as shown. Since a is an integer, the piece I must cover equal amounts of black and white; similarly for pieces II and III. Because R also does the same, we have by subtraction that the piece IV must also cover equal areas of black and white. We shall see, however, that this is impossible.

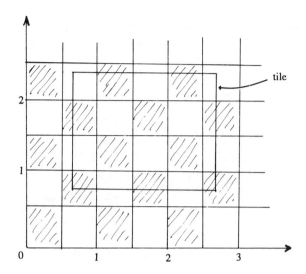

FIGURE 106

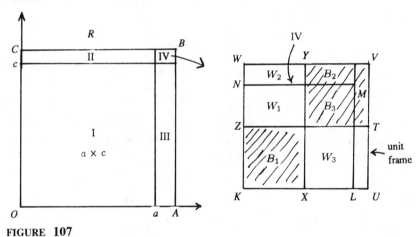

FIGURE 107

The region IV is a piece that is cut from the bottom left corner of a unit square section of the checkerboard, i.e., from a section consisting of 4 squares of the board, 2 black and 2 white. For definiteness, let us suppose the bottom left square is black and that piece IV is $KLMN$.

We can easily show that IV contains more black than white as follows. Let the subregions into which the unit frame $KUVW$ is divided by the edges of IV and the midlines XY and ZT be labelled $B_1, B_2, B_3, W_1, W_2, W_3$ (for black and white), as shown. Then clearly

$$B_1 = W_1 + W_2$$

and

$$B_2 + B_3 = W_3,$$

giving

$$B_1 + B_2 + B_3 = W_1 + W_2 + W_3$$

and

$$B_1 + B_3 = (W_1 + W_3) + (W_2 - B_2).$$

But $B_2 < W_2$ (W_2 is a full $WN \times \frac{1}{2}$, but B_2 isn't as wide), making $W_2 - B_2$ positive. Thus

$$\text{the black part of } IV = B_1 + B_3$$

$$> W_1 + W_3$$

$$= \text{the white part of } IV.$$

The cases $KL \leq \frac{1}{2}$, and $KN \leq \frac{1}{2}$, are similar (and even simpler). Thus IV always covers more black than white, and with this contradiction the argument is complete.

AN AMAZING LOCUS

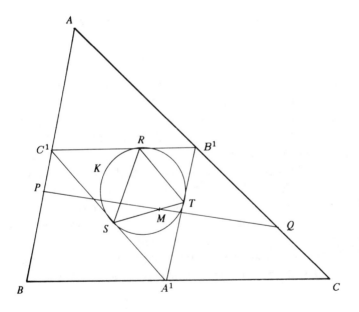

FIGURE 108

ABC is a given triangle and $A'B'C'$ is its *medial* triangle, that is, the triangle determined by the midpoints of the sides. Suppose the incircle K of the medial triangle touches its sides at R, S, and T, as shown. Let P be any point on a side of $\triangle ABC$, and let Q be the point obtained by going halfway around the triangle, so that PQ bisects the perimeter of $\triangle ABC$; there is such a segment, then, for every point P on a side of $\triangle ABC$. Prove the remarkable fact that

the midpoint M of PQ always lies on a side of $\triangle RST$.

Solution

Obviously, as P moves around ABC, Q has to keep its distance and therefore recedes appropriately, causing PQ to vary and M to trace out some locus. The proof divides nicely into two main sections:

 (i) show that the locus passes through each of the vertices R, S, and T,
(ii) show that the locus is piece-wise linear.

   Since part (ii) is much easier to establish, let us begin with it.

   Because ABC has only 3 angles, PQ cannot separate them two on each side; that is to say, on one of the sides of PQ there must be only one angle of $\triangle ABC$, and therefore PQ can always be regarded as moving from position to neighboring position so as to cut off the same total length (namely one-half the perimeter) from the two arms of an angle of $\triangle ABC$ (e.g., angle A in the above illustration). We need only show, then, that as P and Q vary, one on each arm of a given angle $VOW = \theta$, so as to cut a constant total length $(2r + 2s)$ from the two arms, that the midpoint M of PQ travels in a straight path.

   If PQ cuts $2r$ from one arm and $2s$ from the other, the coordinates of P and Q, with coordinate axes assigned as shown below are

$$P(2r\cos\theta,\ 2r\sin\theta) \quad \text{and} \quad Q(2s, 0),$$

giving the midpoint M to be

$$M(s + r\cos\theta,\ r\sin\theta).$$

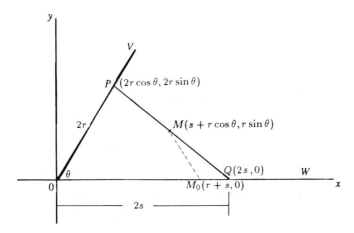

FIGURE 109

Now consider the fixed point $M_0(r+s, 0)$ on the x-axis. The slope of MM_0 is

$$\frac{r \sin \theta}{r \cos \theta - r} = \frac{\sin \theta}{\cos \theta - 1},$$

a constant. Since M_0 is a fixed point and M is an arbitrary point on the locus, it follows that M must generate a *straight* locus. Now let's go on to part (i).

Since there is no preference between the vertices R, S, and T, we need only show that any one of them, say R, lies on the locus on M. Although none of the steps involved is difficult, we shall need to build up to the conclusion in several stages.

First, consider the incircle I of a $\triangle ABC$. Let us adopt the usual notation: the side opposite A has length a, and so on, and the semiperimeter $a + b + c = s$; also, let the equal tangents from the vertices have lengths x, y, z as shown.

Then we have

$$2x + 2y + 2z = \text{the perimeter } 2s,$$

giving $x + y + z = s$. Since $x + z = b$, then $y = s - b$, i.e., $BD = s - b$.

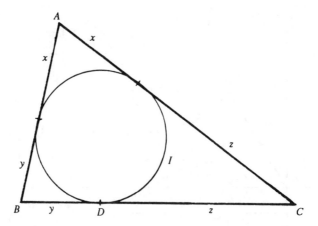

Next, suppose the excircle opposite A touches the sides at U, V, W, as shown, where x and y again mark pairs of equal tangents. Clearly, the perimeter of $\triangle ABC$ can be unfolded at V to cover the equal tangents AU and AW. Thus each of AU and AW must be the semiperimeter s. Because $AC = b$, then y must be $s - b$, making

$$CV = s - b.$$

We note in passing the important fact that AV bisects the perimeter, making AV one of the positions taken by our perimeter-bisecting segment PQ as it varies around $\triangle ABC$.

Putting our results together, we have that, if the incircle and the appropriate excircle touch BC at D and V, then

$$BD = CV \quad (= s - b).$$

If A' is the midpoint of BC, this yields the major result that

$$DA' = A'V.$$

Next let us consider some of the relations between a triangle and its medial triangle $A'B'C'$. Everybody knows that the medians AA', BB', CC' are concurrent at the centroid G and each is divided by G in the

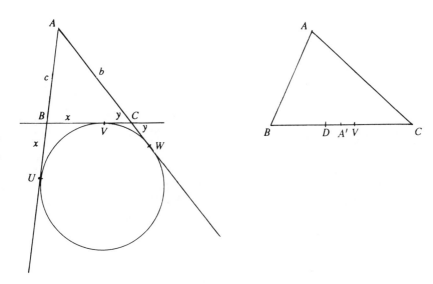

FIGURE 111

ratio 2: 1. Consequently, the dilatation $G(1/2)$ (that is, the transformation which carries each point P straight through G to an image point P'
that is half as far from G on the other side) carries A into A', B into B',
and C into C', that is, it transforms $\triangle ABC$ into $\triangle A'B'C'$. More to the
point, it carries the incircle I of $\triangle ABC$ into the incircle K of $\triangle A'B'C'$
($\triangle A'B'C'$ with its incircle K is merely a scaled- down version of $\triangle ABC$
and its incircle I). In doing so, of course, all the points of BC are carried into the points of $B'C'$; in particular, A' goes to the point A'' (on
$A'A$) and the point of contact D is taken into the point of contact R.
The crucial matter here is that everything in $\triangle A'B'C'$ is scaled down to
half-size from $\triangle ABC$, and we have that the image

$$A''R = \frac{1}{2}DA'.$$

But, since $DA' = A'V$, it follows that

$$A''R = \frac{1}{2}A'V,$$

which leads to the collinearity of A, R, and V as follows.

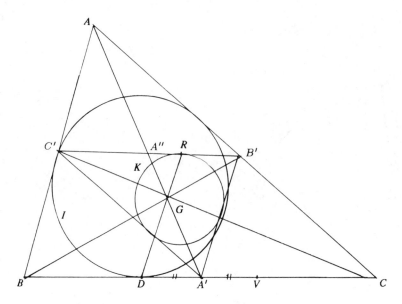

FIGURE 112

It is well known that $B'C'$ is parallel to BC and that it bisects every segment from A to a point in BC, in particular AA' (at A''). Suppose that AR extended crosses BC at some point Z. Clearly then the triangles $AA''R$ and $AA'Z$ are similar and have corresponding sides in the ratio $AA''/AA' = 1/2$. Thus

$$A'Z = 2A''R.$$

Since $A''R = \frac{1}{2}A'V$, then we conclude that

$$A'Z = A'V, \quad \text{and} \quad Z = V.$$

Hence A, R, and V are collinear alright, in which case AV is bisected by $B'C'$ at R. Recalling that AV is indeed a position taken by the variable segment PQ, the midpoint M, upon such an occasion, would be located at R, and part (i) is established.

For a final look at what's going on, suppose the incircle and excircles touch $\triangle ABC$ at the pairs of points (D, V), (E, W), (F, X), as

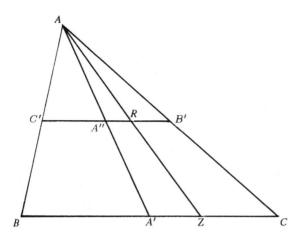

FIGURE 113

shown. If PQ starts in position AV and P moves along AX, Q will proceed along VC to C, and M would trace the side RT of $\triangle RST$. As P continues to B, Q goes along CA to W, while M traverses TS; finally, P completes its trip halfway around by covering BV, as Q traces WA and M goes along SR.

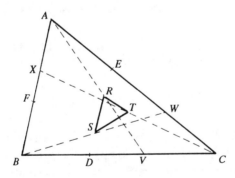

FIGURE 114

MOESSNER'S THEOREM

1. The Theorem

Suppose the positive integers are written in order in a row and that every 5th number is underlined. Ignoring the underlined integers, let the partial sums of the other numbers in the row be recorded in a second row, placing the sum directly beneath the last entry that is contained in the sum.

1	2	3	4	5	6	7	8	9	10	11	12	13	14	15	16	
1	3	6	10		16	23	31	40		51	63	76	90		106	$\cdots$
1	4	10			26	49	80			131	194	270			376	$\cdots$
1	5				31	80				211	405				781	$\cdots$
1					32					243					1024	$\cdots$

In the second row, let every 4th number be underlined and ignored, and the partial sums of the remaining numbers recorded in a third row, again placing the sum under the last term that it includes. In like fashion, proceed to a fourth and a fifth row, ignoring every 3rd number in row 3 and every 2nd number in row 4. Then the entries in the final row turn out to be the perfect fifth powers $1^5, 2^5, 3^5, 4^5, \ldots$!

Had we begun by ignoring every 4th number in the first row, every 3rd number in the second row and every 2nd number in the third row, we should have wound up with the perfect fourth powers in row 4. In fact, for every positive integer $k > 1$, if every kth number is ignored in row 1, every $(k-1)$th number in row 2, and, in general, every $(k+1-i)$th number in row i, then the kth row of partial sums will turn out to be just the perfect kth powers $1^k, 2^k, 3^k, \ldots$.

This was discovered in 1951 by Alfred Moessner, who is internationally known in the field of recreational mathematics for many spectacular results in arithmetic (many of his gems are to be found among the "Curiosa" which were featured regularly in the journal *Scripta Mathematica* during the years 1932–1957). This theme was taken up by various noted mathematicians in the next 15 years and several generalizations were established. Since the proofs of these results are generally encumbered with extensive use of binomial coefficients, it is a pleasure to present a simple new approach that is due to Karel Post, a very clever mathematician at the Eindhoven University of Technology in The Netherlands.

2. Post's Proof

Obviously the gaps in our array produce a saw-tooth pattern of triangular sections, each an isosceles right triangle extending over k rows and k columns. Our first step is to observe that the array can be generated equally well by the above "Moessner procedure" when one begins with an initial row of 1's in which every $(k + 1)$th entry is underlined and ignored; in this case the triangular sections have breadth and depth $k+1$ instead of just k, and the rows are respaced to accommodate these slightly larger sections.

1	1	1	1	1	1̲	1	1	1	1	1	1̲	1	1	···
1	2	3	4	5̲		6	7	8	9	10̲		11	12	···
1	3	6	10̲			16	23	31	40̲			51	63	···
1	4	10̲				26	49	80̲				131	194	···
1	5̲					31	80̲					211	405̲	
1						32						243		···

The Augmented Array; $k = 5$; triangular sections 6×6.

Now Professor Post proposes constructing a graph g from the augmented array by placing a vertex at each of its entries and by inserting an edge between each pair of consecutive vertices in a row or in a column; thus the triangular character of the array is carried over to the graph. The graph is completed by directing each horizontal edge toward the right and each vertical edge downward.

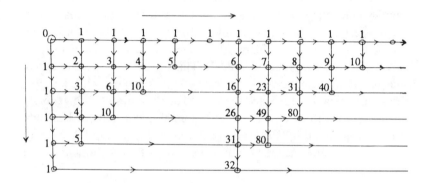

FIGURE **115**

A vertex generally has an edge or two that runs *into* it and an edge or two that runs *out* of it. The upper left corner O is an exception, having no edge that is directed inward. Consequently, let us take this vertex O as an origin and assign to each vertex V a label denoting the *number* of different directed paths from O to V, that is, paths that traverse each edge only in its assigned direction (i.e., to the right or down as the case may be).

Accordingly, each vertex in the upper row is labelled 1, and so is every vertex in the first column; at O itself we arbitrarily assign the label 1.

Since every vertex V can be entered only from the left or from above (the other edges at V are directed *away* from V), the number assigned to V is simply the sum of the label on the vertex to its left and the label on the vertex above it. But this is precisely the rule of partial sums whereby the entries in the original array are determined. We conclude, then, that each number in the array simply denotes the number of directed paths to that position from the corner vertex O. That is to say,

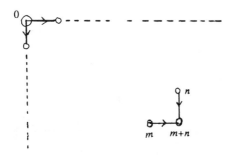

FIGURE 116

in assigning the labels to the vertices in g we are actually reproducing the original array.

Post's next move is to observe that there will be the same number of directed paths in g from O to an arbitrary vertex V as there are from V to O in the graph G that is derived from g by reversing the directions of all its edges—obviously, reversing the directions of all the edges in a directed (O, V)-path converts it into a directed (V, O)-path.

Moessner's theorem, then, claims that there are n^k directed paths in G from the nth vertex V in the bottom row to the upper left corner vertex O. Fortunately, the triangular character of G makes counting these paths a relatively easy undertaking.

Suppose that V is the nth vertex in the bottom row of G. Let each vertex U be labelled with the number of different directed paths from V to U; we wish to prove that the label on vertex O is n^k. Recall that G has $k + 1$ rows; let us number them $0, 1, 2, \ldots, k$ from bottom to top.

Since the edges of G are directed either to the left or upward, the label on a vertex U is the sum of the labels on the vertices to its immediate right and immediately below it.

Now every vertex in the bottom row is the bottom left corner of one of the triangular sections, and so V is at the corner of the nth triangular section, which we shall denote by T_n. Clearly, the labels on the $k+1$ vertices that stand in the column directly above V are all 1's (we arbitrarily assign a label of 1 to V itself). This set of $k + 1$ 1's may be considered to be the $k + 1$ powers $(a^0, a^1, a^2, \ldots, a^k)$, where $a = 1$. The crux of our argument is the proof of the observation that if the labels on the $k + 1$

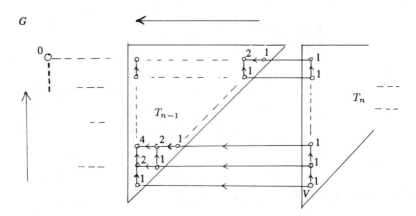

FIGURE 117

vertices in the first column of a triangular section T_r, listed from bottom to top, are the powers of a number a, that is $(a^0, a^1, a^2, \ldots, a^k)$, then the labels on the corresponding vertices in the first column of the next triangle T_{r-1}, which is one stage closer to the corner O, will turn out to be the powers of the next integer, $(a+1)$, that is $((a+1)^0, (a+1)^1, \ldots, (a+1)^k)$. Once this lemma has been established, the conclusion follows easily: We have already noted that this condition *is* satisfied at V in triangle T_n, where the labels are $(1, 1, 1^2, \ldots, 1^k)$; hence the first column of T_{n-1} carries the labels $(1, 2, 2^2, \ldots, 2^k)$, which, in turn, causes the first column of T_{n-2} to have the labels $(1, 3, 3^2, \ldots, 3^k)$, and so on, to the labels $(1, n, n^2, \ldots, n^k)$ for the first column of T_1, implying the label at O is n^k.

It only remains to prove our central lemma.

To this end, suppose that the labels in the first column of triangle T_r, *from bottom to top*, are $(1, a, a^2, \ldots, a^k)$. We observe immediately that this column of labels can be transferred directly to the $k + 1$ vertices in the last diagonal of T_{r-1}, since each vertex on this diagonal has only one incoming edge, namely the one from the corresponding vertex in the first column of T_r. Because each label is just the sum of the one below it and the one to its right, this is all we need in order to determine, one diagonal at a time, all the labels in the triangle T_{r-1}.

If the labels on one diagonal are $(1, a, a^2, \ldots, a^k)$, then the labels on the next diagonal are $(1 + a, a + a^2, a^2 + a^3, \ldots, a^{k-1} + a^k)$, that is

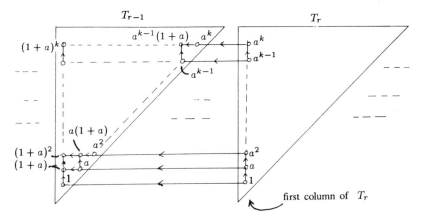

T_{r-1} T_r

FIGURE 118

$(1 + a)$ times the labels in the previous diagonal (except the last one, since the diagonals get shorter by one term at each stage). Similarly, in general, if the labels going along the $(i + 1)$th diagonal are $((1 + a)^i, (1+a)^i a, (1+a)^i a^2, \ldots (1+a)^i a^{k-i})$, then the next one carries labels that are correspondingly $(1 + a)$ times as great. Therefore the first *column* of T_{r-1}, consisting of all of the bottom-left corners of these diagonals, will carry the labels $(1, 1+a, (1+a)^2, \ldots, (1+a)^k)$, as claimed, and our argument is complete.

3. A Generalization

If we begin our array with a row of 1's and let K_1 denote the position of the first number to be underlined, K_2 the number of additional positions along the row to the next underlined entry, and so on, then K_i will be the dimensions of the ith triangular section T_i. In Moessner's theorem the constant sequence $K_i = k+1$ yielded the perfect kth powers in the bottom left corners of the triangular sections T_i. Now let us inquire about the numbers that are produced there by other sequences $\{K_i\}$.

First of all we should note that the above graphical analysis is drastically altered when we try to work our way back toward O from a smaller triangle T_r to a larger one T_{r-1} (in this case, T_r does not supply enough labels to give complete coverage of the last diagonal of T_{r-1}; even though the labels that are lacking all are 0's, our analysis would have

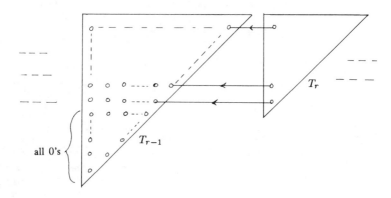

FIGURE 119

to be thoroughly revised to handle such a case, if indeed it would suffice to solve the problem.

Accordingly, let us consider only nondecreasing sequences $\{K_i\}$; in this case, the labels in T_{r-1} are easily determined from the first column of T_r, even though the $K_r - K_{r-1}$ lowest labels are not used directly.

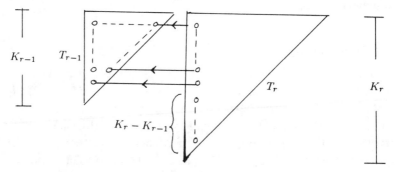

FIGURE 120

For example, consider the array A constructed on the basis that $K_i = i + 1$. The first few results strongly suggest that it is the factorials that are being generated; we shall see that this is indeed the case.

$A:$

1	1	1	1	1	1	1	1	1	1	1	1	1	1	1	$\cdots$
1		2	3		4	5	6		7	8	9	10		11	$\cdots$
		2			6	11			18	26	35			46	$\cdots$
					6				24	50				96	$\cdots$
									24					120	$\cdots$
														120	$\cdots$

In the general case of an arbitrary nondecreasing sequence $\{K_i\}$, the entry at the vertex V in the corresponding graph G which is at the bottom left corner of the triangular section T_n is still just the number of directed (V, O)-paths to the upper left corner O. Again, the labels on the vertices in the column that stands on V are all 1's and constitute the K_n-tuple $(1, 1, 1^2, \ldots, 1^{K_n-1})$. By an analysis that exactly parallels the previous proof, we can see that if the labels in the first column of T_r are a constant t times the powers of $n + 1 - r$ (let $n + 1 - r = a$ for short), that is $(t \cdot 1, ta, ta^2, \ldots, ta^{K_r-1})$, then the K_{r-1} labels in the first column of T_{r-1} are a constant s times powers of the next integer, $(a+1)$, that is $(s \cdot 1, s(a+1), s(a+1)^2, \ldots, s(a+1)^{K_{r-1}-1})$, where $s = t \cdot a^{K_r-K_{r-1}}$. It's fairly easy to see from a figure.

Accordingly, comparing the label q_{r-1} in the upper left corner of T_{r-1} and the label q_r in the upper left corner of T_r, we obtain the formula

$$q_{r-1} = q_r \cdot \frac{(a+1)^{K_{r-1}-1}}{a^{K_{r-1}-1}}.$$

In these terms, the desired label at O is q_1. From the obvious $q_n = 1$, we have

$$q_{n-1} = 2^{K_{n-1}-1};$$

$$q_{n-2} = 2^{K_{n-1}-1} \cdot \frac{3^{K_{n-2}-1}}{2^{K_{n-2}-1}} = 2^{K_{n-1}-K_{n-2}} \cdot 3^{K_{n-2}-1};$$

$$q_{n-3} = q_{n-2} \cdot \frac{4^{K_{n-3}-1}}{3^{K_{n-3}-1}} = 2^{K_{n-1}-K_{n-2}} \cdot 3^{K_{n-2}-1} \cdot \frac{4^{K_{n-3}-1}}{3^{K_{n-3}-1}}$$

$$= 2^{K_{n-1}-K_{n-2}} \cdot 3^{K_{n-2}-K_{n-3}} \cdot 4^{K_{n-3}-1}.$$

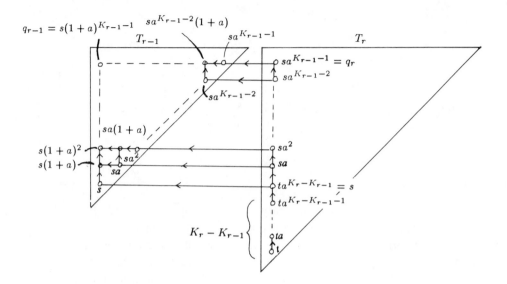

FIGURE 121

and so on, to the general formula

$$q_1 = 1^{K_n - K_{n-1}} \cdot 2^{K_{n-1} - K_{n-2}} \cdots (n-1)^{K_2 - K_1} \cdot n^{K_1 - 1}, \quad \text{i.e.,}$$

$$q_1 = \prod_{i=1}^{n} i^{K_{n+1-i} - K_{n-i}}, \quad \text{where} \quad K_0 = 1.$$

In Moessner's theorem we have $K_i = k + 1$ for all i except K_0, which equals 1 by definition, making $K_{i+1} - K_i = 0$ for all $i > 0$, and the number at V to be

$$q_1 = 1^0 \cdot 2^0 \cdot 3^0 \cdots (n-1)^0 \cdot n^{K_1 - K_0} = n^{k+1-1} = n^k.$$

Also, in the later case of $K_i = i + 1$, we have $K_{i+1} - K_i = 1$ for all i, including $i = 0$, yielding the number at V to be

$$q_1 = \prod_{i=1}^{n} i^1 = n!,$$

as noted.

Obviously there are many variations that can be played on this theme— using different initial rows instead of a row of 1's (e.g., some arithmetic progression); perhaps an occasional hour could be passed with pleasure and profit in such an enterprise.

REFERENCES

1. Guy, R.K., *Reviews in Number Theory 1973–1983*, Vol. 1A, Section B60, (1984) 319–320, American Mathematical Society.
2. Leveque, W.J., *Reviews in Number Theory*, as printed in *Mathematical Reviews*, 1940 through 1972, Vol. 1, Section B60 (1974) 335–339, American Mathematical Society.
3. Long, C.T., Strike it out—add it up, *The Mathematical Gazette*, 66 (1982) 273–277.
4. Paasche, I., Eine Verallgemeinerung des Moessnerschen Stazes, *Compositio Math*, 12 (1956) 263–270.
5. Slater, J.G., Strike it out—some exercises, *The Mathematical Gazette*, 67 (1983) 288–290.
6. Yzeren, J. van, A note on an additive property of natural numbers, *The American Mathematical Monthly* (1959) 53–54.

COUNTING TRIANGLES

Three rods of lengths a, b, c can be made to form a triangle if and only if the three triangle inequalities are satisfied:

$$a + b > c, \quad b + c > a, \quad c + a > b.$$

Suppose one wants to make just one triangle and has at his disposal exactly one rod of each of the lengths $1, 2, 3, \ldots, n$. He might decide on the (2,4,5) triangle, the (3,4,5), or the (3, 5, 7), …. He can't make an isosceles triangle, of course, but there are still many possibilities to choose from; the question is: How many?

Solution

Our approach will be by the well travelled path of *recursion*. While this will guide us toward certain general goals, one need not be concerned that we shall be occupied with just a mechanical application of a formal procedure; on the contrary, it is clever thinking and the skillful handling of our resources that will occupy us throughout.

Let $T(n)$ be the required number of individual triangles that can be made from a stock of one rod of each of the lengths $1, 2, 3, \ldots, n$. If a new rod, of length $n + 1$, were to be added to our stock, some new possibilities would arise and the total number of constructible triangles would increase to $T(n + 1)$. The number of new triangles would be

$T(n + 1) - T(n)$, and in each new triangle the new rod would have to occur. But we can count these new triangles, for the other two sides, a and b, must add up to more than $n + 1$, and conversely; that is to say, $(a, b, n + 1)$, where $a, b \in \{1, 2, 3, \ldots, n\}$, constitutes an acceptable triangle if and only if $a + b > n + 1$. Thus our problem reduces to that of counting the suitable pairs (a, b).

The range of values of $a + b$ runs from a minimum of $n + 2$ to a maximum of $n + (n - 1) = 2n - 1$, and it includes every integer in between. In general there are several combinations that give the sum $n + 2$; we might choose

$$(2, n), \ (3, n - 1), \ (4, n - 2), \ldots .$$

In order to complete the list of cases, one must consider whether n is odd or even; similarly for $a + b = n + 3$, and the larger sums. However, if we *start at the other end*, with sum $2n - 1$, there is always only one choice for a and b, namely $(n, n - 1)$. Similarly, there is only one choice in the case of the sum $2n - 2$: $(n, n - 2)$ (recall that $n - 1$ cannot be repeated). Continuing, we find two solutions for the sum $2n - 3$: $(n, n - 3)$ and $(n - 1, n - 2)$; and two more for the sum $2n - 4$: $(n, n - 4)$, and $(n - 1, n - 3)$. To provide a sum of $2n - k$, we cannot dip below $n - k$ for either a or b, for the biggest the other can be is only n. We must choose from the $k + 1$ lengths $S = \{n, n - 1, \ldots, n - k + 1, n - k\}$, and once either a or b has been selected there is no choice left for the other. Consequently, the values (a, b) are obtained simply by pairing up the numbers in S from the outside in:

$$(n, n - k), \ (n - 1, n - k + 1), \ (n - 2, n - k + 2), \ldots .$$

When $k + 1$ is even, say $2t$, this gives t pairs (a, b). The next case, then, has $2t + 1$ lengths to choose from, but there are still only t pairs (a, b), with an unusable length in the middle of S left over. The following case has $2t + 2$ lengths, giving $t + 1$ complete pairs (a, b), and so on. Consequently, for the $n - 2$ sums $2n - 1, 2n - 2, \ldots, n + 2$, the numbers of pairs (a, b) are, respectively,

$$1, 1, 2, 2, 3, 3, \ldots \text{ to } n - 2 \text{ terms,}$$

and we have

$$T(n+1) - T(n) = 1 + 1 + 2 + 2 + 3 + \cdots \text{ to } n-2 \text{ terms.}$$

It remains to solve this recursion for $T(n)$. The cases of n odd and n even are most easily handled separately. However, to do so presents us with a minor problem, for the recursion relates cases concerning *consecutive* values of n. Our final hurdle, then, is to convert the recursion into one that connects either consecutive even values or consecutive odd values of n.

For n even $(n = 2t)$, the recursion yields

$$\begin{aligned}
T(n+1) - T(n) &= 1 + 1 + 2 + 2 + \cdots + (t-1) + (t-1) \\
&= 2 \cdot \frac{[(t-1)t]}{2} = \frac{(2t-2)(2t)}{4} \\
&= \frac{(n-2)n}{4}.
\end{aligned} \tag{1}$$

For n odd $(n = 2t + 1)$, we have

$$\begin{aligned}
T(n+1) - T(n) &= 1 + 1 + 2 + 2 + \cdots + (t-1) + (t-1) + t \\
&= (t-1)t + t = t^2 = \frac{(n-1)^2}{4}.
\end{aligned} \tag{2}$$

Thus we are in possession of the value $T(n+1) - T(n)$, provided the *parity* of n is specified. Consequently, we can derive a recursion relating consecutive even values of n from the following simple but perceptive observation:

$$T(2r) - T(2r-2) = [T(2r) - T(2r-1)] + [T(2r-1) - T(2r-2)].$$

For $n = 2r - 1$, we have, by relation (2) above, that the first term on the right side of this equation is

$$\begin{aligned}
T(n+1) - T(n) &= T(2r) - T(2r-1) \\
&= \frac{(2r-2)^2}{4} = (r-1)^2;
\end{aligned}$$

and *for* $n = 2r - 2$ we have, by relation (1), that the second term is

$$T(n + 1) - T(n) = T(2r - 1) - T(2r - 2)$$
$$= \frac{(2r - 4)(2r - 2)}{4}$$
$$= (r - 1)(r - 2).$$

Hence

$$T(2r) - T(2r - 2) = (r - 1)^2 + (r - 1)(r - 2) = 2r^2 - 5r + 3.$$

Using this repeatedly, we obtain

$$
\begin{array}{lllll}
T(2r) & -T(2r - 2) & = 2r^2 & -5r & +3 \\
T(2r - 2) & -T(2r - 4) & = 2(r - 1)^2 & -5(r - 1) & +3 \\
T(2r - 4) & -T(2r - 6) & = 2(r - 2)^2 & -5(r - 2) & +3 \\
& & \cdots & & \\
T(6) & -T(4) & = 2(3)^2 & -5(3) & +3.
\end{array}
$$

Adding gives

$$T(2r) - T(4) = 2\sum_{k=3}^{r} k^2 - 5\sum_{k=3}^{r} k + 3(r - 2)$$

$$= 2\sum_{k=1}^{r} k^2 - 5\sum_{k=1}^{r} k + 3(r - 2) - 2(1^2 + 2^2) + 5(1 + 2).$$

For $n = 4$, we have only one possible triangle: (2,3,4) (none of the other three triples from $\{1, 2, 3, 4\}$ determine a triangle—(1,2,3), (1,2,4), and (1,3,4)). Hence $T(4) = 1$, and we get

$$T(2r) = 1 + \frac{2r(r + 1)(2r + 1)}{6} - \frac{5r(r + 1)}{2} + 3r - 6 - 10 + 15,$$

which simplifies in a straightforward manner to

$$T(2r) = \frac{2r(2r - 2)(4r - 5)}{24}.$$

Therefore, *for n even* $(n = 2r)$, we have the result

$$T(n) = \frac{n(n - 2)(2n - 5)}{24}.$$

Finally, knowing T (even), we can deduce T (odd) from the original recursion. For n even we have above that

$$T(n+1) - T(n) = \frac{n(n-2)}{4};$$

hence

$$T(2r+1) - T(2r) = \frac{2r(2r-2)}{4} = r(r-1),$$

and

$$\begin{aligned}
T(2r+1) &= T(2r) + r(r-1) \\
&= \frac{2r(2r-2)(4r-5)}{24} + \frac{6[2r(2r-2)]}{24} \\
&= \frac{2r(2r-2)(4r+1)}{24}.
\end{aligned}$$

For n odd, then, $(n = 2r+1)$, we have

$$T(n) = \frac{(n-1)(n-3)(2n-1)}{24}.$$

A GEOMETRIC MINIMUM

Let the fixed point P be taken anywhere inside the lens-shaped region of intersection R of two given circles C_1 and C_2. Let UV be a chord of R through P. Determine how to construct the chord which makes the product $PU \cdot PV$ a minimum.

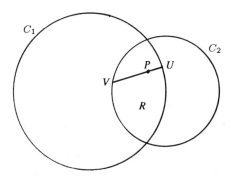

FIGURE 122

Solution

Let C_1 be the circle $O_1(r_1)$, that is, with center O_1 and radius r_1, and let C_2 be $O_2(r_2)$ (see Figure 123). Clearly, by symmetry, the external common tangents to the circles meet at some point H on the line of centers O_1O_2 (H is often a point of major importance and is very popular with geometers; for example, H is the center of a dilatation which transforms C_1 into C_2). It turns out that HP determines the desired chord.

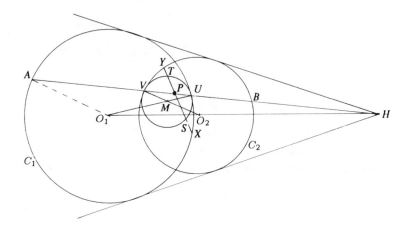

FIGURE 123

Let HP meet the circles again at A and B, as shown; also, let O_1U cross O_2V at M. Now H is the center of a dilatation, with ratio HO_1/HO_2, which carries C_2 into C_1, in particular taking O_2 into O_1 and V into A. Since a dilatation does not change the direction of a line, this means that AO_1 and VO_2 are parallel, and we have equal corresponding angles O_1AV and O_2VU. But $\triangle O_1AU$ is isosceles ($O_1A = O_1U = r_1$), making $\angle O_1AU = \angle O_1UA$. Thus

$$\angle O_2VU = \angle O_1UA,$$

and $\triangle MUV$ is isosceles with $MU = MV$.

Consequently the circle $C = M(MU)$ passes through U and V and, in fact, is internally tangent to each of C_1 and C_2 (for example, in the case of C_1, the distance $O_1 M$ between their centers is equal to the difference $O_1 U - MU$ of their radii). That is to say, this circle C, being tangent to R at U and V, is contained completely within R, and any chord of R through P, other than UPV, must extend beyond C to points X and Y on the boundary of R. If XY crosses C at S and T, then we have

$$PX \cdot PY \geq PS \cdot PT = PU \cdot PV \quad \text{in} \quad C,$$

yielding the desired conclusion.

Furthermore, consider a variable chord through P that extends to cross C_1 (say at A and U) and C_2 (at B and V). As the chord varies through P, the product $PA \cdot PU$ remains constant (in C_1), as does the product $PV \cdot PB$ (in C_2). Accordingly, for all positions, the combined product

$$PA \cdot PU \cdot PV \cdot PB \quad \text{remains constant.}$$

As a little bonus, then, we have the corollary that the chord HP, in making $PU \cdot PV$ a minimum, also makes $PA \cdot PB$ a maximum!

THE PROBABILISTIC METHOD

Suppose you wanted to prove that there is a map that cannot be colored with just 4 colors. Although you might not be able to produce one, it would be sufficient for your purpose, if you were good enough at counting maps, to be able to show that the *number* of such maps is *positive*. In the language of probability, this alternative approach would amount to the problem of showing that the probability p that a randomly chosen map would require more than 4 colors is a *positive* number. At the mention of probabilities, one might feel some uneasiness, for probabilities are irrevocably bound up with risk and uncertainty. There is no denying that success at showing the probability p is positive would not remove the risk, in an actual experiment, of failing to have a random selection turn up a map of the prescribed kind; such an enterprise would still be highly uncertain.

But we are not interested in performing any experiment. All we want out of this is that the probability p is positive, and if there is no doubt about the *proof* of that, then there is nothing at all to be uncertain about; surely if no map exists which requires more than 4 colors, it would be impossible to really *prove* that there is a positive probability of selecting such a map.

Such is the basis of "the probabilistic method." In the last 30 years or so, the success of this approach has been both widespread and significant. A wealth of applications is given in the monograph [1], and the example that follows, due to the illustrious Paul Erdős, although appearing in [1] (page 40), has been taken primarily from the excellent expository article "Combinatorics by Coin Flipping" by Joel Spencer (State University of New York at Stony Brook), in the *College Mathematics Journal*, November 1986, pages 407–412.

Tournaments and the Property S_k

If each of the $\binom{n}{2}$ pairs from a set of n vertices, $(1, 2, \ldots, n)$, is joined by a directed arc, a "tournament" T_n is determined. The results of an ordinary round-robin tournament with n contestants determine a T_n if the directions are assigned to the arcs ij so as to point toward the loser of the match between players i and j. Since each arc may be directed in either of two possible ways, there are $2^{\binom{n}{2}}$ different tournaments T_n of "order" n (i.e., on n vertices).

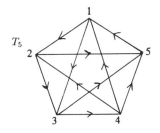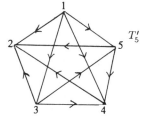

FIGURE 124

If k is a positive integer not exceeding n, the vertices in a T_n would determine $\binom{n}{k}$ subsets of size k. A tournament T_n is said to have property S_k if, for *each* subset X having k vertices, there is some player, *not in X*, who won all k of his matches against the members of X (that is, somebody *dominates X*). Thus the tournament T_5 above has property S_1, while T_5' does not (vertex 1 spoils things). The T_7 shown below has property S_2, that is, for each of the $\binom{7}{2} = 21$ pairs of vertices, somebody beat them both (e.g., 5 and 6 both lost to 7; 3 and 7 both lost to 4; ...).

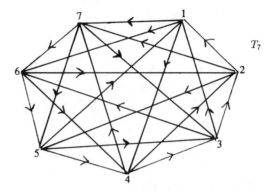

T_7

FIGURE 125

It has been shown that there is no tournament having property S_3 which has fewer than 31 vertices, and so it is not a simple matter to illustrate property S_3. For S_4, T_n cannot have fewer than 79 vertices, and for S_5, not fewer than 191. Before devoting any substantial amount of work to the task of constructing tournaments with properties S_3, S_4, S_5, etc., it would be nice to be sure that they are there to be found. Our problem, then, is the following:

> Does there exist a tournament T_n having property S_k for each positive integer k?

Solution

Paul Erdős was able to show by a beautiful and simple argument that, in a certain collection of tournaments, the probability of a random selection having property S_k is a positive number.

　　Let k be specified in advance and let n be an arbitrary positive integer. We have seen above that, for given n, there exists a set T of $2^{\binom{n}{2}}$ different tournaments T_n with n vertices. We propose to show that there exists, for some positive integer n, such a set of tournaments T for which there is a positive probability that a randomly selected T_n has property S_k. We shall select a tournament from T by constructing one arc-for-arc as follows. Beginning with the underlying undirected complete graph on

n vertices, we decide on the direction to give an arc by flipping a fair coin for each arc (further details are inessential).

This certainly generates a T_n, and all tournaments T_n are equally likely. It also has the crucial property that the probability that a particular arc points in a particular direction is $\frac{1}{2}$ in *all* cases. On this basis we can easily deduce the probability that a tournament T_n, constructed randomly in this way, has property S_k. Actually we shall deduce the complementary probability that it fails to have property S_k, and subtract it from 1.

Consider any subset X of k vertices of T_n. For each vertex v not in X, the probability that all k of its arcs to X are directed toward X (i.e., that v beat everybody in X), is $(\frac{1}{2})^k = 2^{-k}$ (see Figure 126). We can think of this as the probability that v "supports" the property S_k with respect to the subset X. The complementary probability that v fails to support property S_k is therefore $1 - 2^{-k}$.

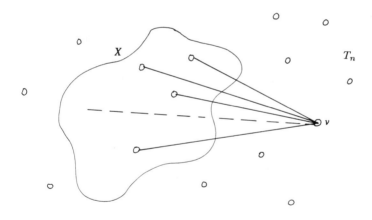

FIGURE 126

The probability that all $n - k$ of the vertices v that lie outside X fail to support property S_k, with respect to X, is $(1 - 2^{-k})^{n-k}$. That is to say, the probability that the subset X spoils property S_k is $(1 - 2^{-k})^{n-k}$. Now, the failure of any one of the $\binom{n}{k}$ possible subsets X with k vertices would be enough to spoil property S_k. Of course, there is a great deal of overlapping of these subsets. However, the probability that at least one of them denies property S_k could not possibly be more than the sum

$$(1 - 2^{-k})^{n-k} + (1 - 2^{-k})^{n-k} + \cdots = \binom{n}{k} \cdot (1 - 2^{-k})^{n-k}.$$

The probability that this does *not* happen, that is, that T_n *has* property S_k because no subset spoils it, is then *at least*

$$p(n, k) = 1 - \binom{n}{k} \cdot (1 - 2^{-k})^{n-k}.$$

It only remains to show that, for every k, some n is big enough to make $p(n, k)$ a positive number. We don't care how very large n might have to be, just that some such n exists. If we could show that the variable term

$$f(n, k) = \binom{n}{k} \cdot (1 - 2^{-k})^{n-k}$$

is ever less than 1, the desired conclusion would follow.

To this end, consider the ratio

$$
\begin{aligned}
\frac{f(n+1, k)}{f(n, k)} &= \frac{\binom{n+1}{k}(1 - 2^{-k})^{n+1-k}}{\binom{n}{k}(1 - 2^{-k})^{n-k}} \\
&= \frac{(n+1)! k! (n-k)! (1 - 2^{-k})}{k! (n+1-k)! n!} \\
&= \frac{(n+1)(1 - 2^{-k})}{n+1-k} \\
&= \frac{1 - \frac{1}{2^k}}{1 - \frac{k}{n+1}},
\end{aligned}
$$

implying

$$f(n+1, k) = \frac{1 - \frac{1}{2^k}}{1 - \frac{k}{n+1}} \cdot f(n, k).$$

Similarly,

$$f(n+2, k) = \frac{1 - \frac{1}{2^k}}{1 - \frac{k}{n+2}} \cdot f(n+1, k).$$

Putting $k/(n+1)$ for $k/(n+2)$ on the right side gives

$$f(n+2, k) < \frac{1 - \frac{1}{2^k}}{1 - \frac{k}{n+1}} f(n+1, k)$$

$$= \left\{ \frac{1 - \frac{1}{2^k}}{1 - \frac{k}{n+1}} \right\}^2 \cdot f(n, k),$$

and eventually, for any positive integer $r > 1$,

$$f(n+r, k) < \left\{ \frac{1 - \frac{1}{2^k}}{1 - \frac{k}{n+1}} \right\}^r \cdot f(n, k).$$

Now, for some $n = N$, we must have $k/(N+1) < 1/2^k$, giving

$$c = \frac{1 - \frac{1}{2^k}}{1 - \frac{k}{N+1}} < 1, \qquad c \text{ a } constant.$$

Hence, for all $n > N + 1$, i.e., for $n = N + r, r > 1$,

$$f(n, k) = f(N + r, k) < c^r \cdot f(N, k).$$

As n grows without bound, so does r, and since $c < 1$ and $f(N, k)$ is a constant, we have

$$c^r \cdot f(N, k) \to 0, \quad \text{forcing} \quad f(n, k) \to 0.$$

Thus $f(n, k)$ is certainly less than 1 for some integer n, and the proof is complete.

We note in passing that as $f(n, k) \to 0$, $p(n, k)$ gets close to 1, and stays close to 1. That is to say, as n gets large, almost every T_n has property S_k!

AN APPLICATION
OF GENERATING FUNCTIONS

In October 1983, *Crux Mathematicorum* carried the following engaging problem, proposed by Leroy Meyers of Ohio State University (Problem 87, page 242):

> How many 9-digit positive integers are there in which no digit occurs 3 or more times consecutively? (The integers are allowed to begin with 0's, but not more than 3 of them, of course.)

The following is not one of the 4 excellent solutions that was published in the 1985 volume (pages 23–25 and 223); although I personally think it is a brilliant approach, no claim to superiority is made for it, for the published solutions are obviously very fine, indeed.

Solution

To begin, consider all the sequences, including the empty sequence ϵ, that can be composed from a given alphabet $(x_1, x_2, \ldots, x_k)$:

$$\epsilon, x_1, x_2, \ldots, x_k, x_1 x_1, x_1 x_2, \ldots.$$

Let A be the set of all such sequences of all lengths and let its generating function be $A(x_1, x_2, \ldots, x_k)$. If we use the same symbol x_i to serve for both the symbol and its algebraic representation, the generating function is merely the sum of all the sequence-symbols themselves, with the empty sequence ϵ represented by $x^0 = 1$:

$$A(x_1, x_2, \ldots, x_k) = 1 + x_1 + x_2 + \cdots + x_1 x_1 + \cdots$$
$$= \sum x_i x_j \ldots,$$

taken over all sequences of all lengths. Collecting terms of the same degree, we have

$$A(x_1, x_2, \ldots, x_k) = 1 + (x_1 + x_2 + \cdots + x_k)$$
$$+ (x_1 + x_2 + \cdots + x_k)^2 + \cdots$$
$$= [1 - (x_1 + x_2 + \cdots + x_k)]^{-1}$$

(just as $1 + z + z^2 + z^3 + \cdots = (1 - z)^{-1}$).

Next, let H be the set of all sequences that can be composed from the alphabet $(x_1, x_2, \ldots, x_k)$ in which *adjacent elements are distinct* (i.e., no *consecutive* repetitions at all). This is not the set we are really interested in, but it's close and is fundamental to the derivation of the set in question. Let's work out its generating function $H(x_1, x_2, \ldots, x_k)$.

The sequences of A and H are related in an obvious way: if the consecutive strings of repeated digits in a member X of A are each contracted to a single digit (i.e. just omit the consecutive duplications), a member Y of H is obtained: e.g.,

$$X = \quad 6 \quad 444 \quad 0 \quad 33 \quad 000000 \quad 52 \quad 99 \quad 66 \quad 8 \quad 777777 \quad \text{in } A$$
yields
$$Y = \quad 6 \quad 4 \quad 0 \quad 3 \quad 0 \quad 52 \quad 9 \quad 6 \quad 8 \quad 7 \quad \text{in } H$$

By such contractions, each member X of A is transformed into a unique correspondent Y of H. And conversely, in order to obtain a particular sequence X of A by reversing this process, there is no choice in the way one must proceed (e.g., the X above can only be obtained by starting with the Y above and tripling the 4, doubling the 3, and so forth). Of course, treating a sequence Y from H in different ways results in different members of A; and each Y can be altered in an infinity of ways. In fact, *each symbol x_i in Y can be replaced by a repeating string $x_i x_i \cdots$ of any length whatsoever*. That is to say, in the construction of a sequence Y of H, if, instead of offering you the single choice of the string of length one, namely x_i itself (in H, adjacent elements must be distinct), you were offered all the strings x_i, $x_i x_i$, $x_i x_i x_i$, ... of all lengths, then, instead of being able to construct just the members Y of H, you would be able to construct the entire set A. Each member of A would be produced precisely once in this process. Correspondingly, in the generating function $H(x_1, x_2, \ldots, x_k)$, if each x_i were replaced by

$$x_i + x_i^2 + x_i^3 + \cdots,$$

i.e., by

$$x_i(1 + x_i + x_i^2 + \cdots) = x_i(1 - x_i)^{-1},$$

the result would contain the algebraic representation of every member of A precisely once, making it in fact the generating function $A(x_1, \ldots, x_k)$. Hence we have

$$H(x_1(1 - x_1)^{-1}, x_2(1 - x_2)^{-1}, \ldots, x_k(1 - x_k)^{-1})$$

$$= A(x_1, x_2, \ldots, x_k) = [1 - (x_1 + x_2 + \cdots + x_k)]^{-1}.$$

In order to make the generating function H more manageable, let us effect a change of variables by setting

$$x_i(1 - x_i)^{-1} = y_i.$$

In this case,

$$x_i = y_i(1 - x_i) = y_i - x_i y_i,$$

and

$$x_i = \frac{y_i}{1 + y_i} = y_i(1 + y_i)^{-1}.$$

Then

$$H(y_1, y_2, \ldots, y_k) = \left[1 - \sum_{i=1}^{k} [y_i(1 + y_i)^{-1}]\right]^{-1}.$$

Since we are accustomed to x's, let's change y_i back to x_i to give, finally,

$$H(x_1, x_2, \ldots, x_k) = \left[1 - \sum_{i=1}^{k} [x_i(1 + x_i)^{-1}]\right]^{-1}.$$

Now let S be the desired set of sequences from the alphabet $(x_1, x_2, \ldots, x_k)$ in which no letter occurs 3 or more times consecutively. We have witnessed that the unrestricted manipulation of the set H yields the entire set A, which contains in addition to S all the unwanted sequences with the forbidden strings of repetitions. But *we control* how H shall be manipulated! Recall that in $H(x_1, x_2, \ldots, x_k)$ no term contains any consecutively repeated symbol. If we permit only repeated strings of length 1 and 2, only the desired set S will be generated. That is to say, if we allow the x_i in $H(x_1, x_2, \ldots, x_k)$ to be replaced by only $x_i + x_i^2$ (instead of the entire series $x_i + x_i^2 + x_i^3 + \cdots$), the resulting generating function will be just $S(x_1, x_2, \ldots, x_k)$:

$$H(x_1 + x_1^2, x_2 + x_2^2, \ldots, x_k + x_k^2) = S(x_1, x_2, \ldots, x_k).$$

Accordingly,

$$S(x_1, x_2, \ldots, x_k) = \left[1 - \sum_{i=1}^{k} [(x_i + x_i^2)(1 + x_i + x_i^2)^{-1}] \right]^{-1}.$$

There is no denying that this looks terribly forbidding! However, don't despair, for the complications soon evaporate with the utmost ease; this is part of the charm of the method.

Since there are 10 digits in our alphabet $(0, 1, 2, \ldots, 9)$, we can begin by setting $k = 10$. Now, each sequence of S is represented in its generating function by an exact algebraic replica, the sequences of length n being represented by the terms of *degree* n. Since we are concerned only with 9-digit integers (of length 9), our only interest is to determine the number of terms of degree 9 in $S(x_1, x_2, \ldots, x_k)$. We don't care whether such a term is

$$x_1 x_1 x_3 x_5 x_1 x_7 x_7 x_8 x_4 \quad \text{or} \quad x_2 x_3 x_1 x_9 x_7 x_6 x_6 x_2 x_2;$$

they each count 1 toward the desired total T. In fact, we no longer care to distinguish between the digits of our alphabet; if each digit were to be changed to x, each term of degree 9 would just become x^9, and no other terms would. That is, the desired total T is simply the coefficient of x^9 in the altered generating function $S(x, x, x, x, x, x, x, x, x)$. If we denote the coefficient of x^n in the function $f(x)$ by $[x^n]f(x)$, then we have

$$T = [x^9]S(x, x, \ldots, x)$$

$$= [x^9]\left[1 - \sum_{i=1}^{10}[(x + x^2)(1 + x + x^2)^{-1}]\right]^{-1}$$

$$= [x^9][1 - 10(x + x^2)(1 + x + x^2)^{-1}]^{-1}$$

$$= [x^9]\frac{1}{1 - \frac{10(x+x^2)}{1+x+x^2}}$$

$$= [x^9]\frac{1 + x + x^2}{1 - 9x - 9x^2}.$$

The easiest way to determine this individual coefficient is by recursion. Let

$$\frac{1 + x + x^2}{1 - 9x - 9x^2} = c_0 + c_1 x + c_2 x^2 + c_3 x^3 + \cdots.$$

Then

$$1 + x + x^2 = (1 - 9x - 9x^2)(c_0 + c_1 x + c_2 x^2 + \cdots).$$

Equating coefficients, we get

$$c_0 = 1, \quad c_1 - 9c_0 = 1, \quad \text{giving} \quad c_1 = 10,$$

$$c_2 - 9c_1 - 9c_0 = 1, \quad \text{giving} \quad c_2 = 100,$$

and, for $n > 2$, $c_n - 9c_{n-1} - 9c_{n-2} = 0$, i.e.,

$$c_n = 9c_{n-1} + 9c_{n-2}.$$

Hence

$$c_3 = 990, \quad c_4 = 9810, \quad c_5 = 97200,$$

$$c_6 = 963090, \quad c_7 = 9542610, \quad c_8 = 945513300,$$

and finally the desired

$$c_9 = \mathbf{936845190}.$$

In addition to this solution by generating functions, I offer, for those who might enjoy a more traditional approach, the following beautiful *maximal block* analysis that comprises the solution of Dag Jonsson of Uppsala, Sweden (1985, p. 223).

Any integer naturally decomposes into its consecutive strings, or blocks, of repeated digits. For example,

$$15555700226221222 \quad \text{gives} \quad 1 \ 5555 \ 7 \ 00 \ 22 \ 6 \ 22 \ 1 \ 222.$$

Since blocks of length 3 or more are forbidden, the blocks in our 9-digit integers will all have length 1 (single digits) or 2 (doublets). We should observe in general that a block of length 7, for example, is not to be construed as two abutting blocks of lengths 3 and 4, but strictly as a single block of length 7. The blocks in our decomposition are *maximal* strings; that is, you know you have reached the end of a block when you hit a different digit or come to the end of the integer itself. This automatically implies the fundamental property that *adjacent* blocks have *different* digits.

In a 9-digit integer there isn't room for more than 4 blocks of length 2. We propose to count the integers having exactly k such doublets and add up as k goes from 0 to 4. To explain the analysis, let us look at the specific case of $k = 2$. Such an integer might be

$$4 \ 6 \ 77 \ 0 \ 1 \ 00 \ 7.$$

Every 9-digit integer with 2 doublets would also have 5 single-digit blocks, for a total of 7 blocks altogether. Such an integer can be constructed, then, by selecting from a row of 7 empty blocks, some 2 to be the doublets, and then proceeding along the row entering the digits themselves. Clearly, the 2 blocks which will be the doublets can be chosen in $\binom{7}{2}$ ways. Now, there are 10 choices for the digit which is to go into the first empty block (the *length* of a block has no bearing on our choice of digit to go into it) but, because adjacent blocks must have different digits, there are only 9 choices for each of the remaining 6 blocks. Thus the number of 9-digit integers with 2 doublets is

$$\binom{7}{2} \cdot 10 \cdot 9^6.$$

In the case of k doublets, the total number of blocks would be $9 - k$ (the k doublets and $9 - 2k$ singletons); a selection of k places for the doublets could be made in $\binom{9-k}{k}$ ways. Since the digits could be entered in $10 \cdot 9^{8-k}$ ways (10 for the first block and 9 for each of the remaining blocks), the number of such integers is

$$\binom{9 - k}{k} \cdot 10 \cdot 9^{8-k}.$$

Adding as k runs from 0 to 4, the desired grand total is

$$\sum_{k=0}^{4} \binom{9 - k}{k} \cdot 10 \cdot 9^{8-k} = 936845190.$$

ON RAT-FREE SETS

1.   Suppose you were given a finite set of n points in the plane. Any particular 3 of these points may or may not be the vertices of a right-angled triangle (RAT for short). If, for example, the n points all lie along a straight line, then none of the triangles they determine would be right-angled, and we would say that the set is "RAT-free." There is no difficulty in constructing RAT-free sets that aren't all collinear; a given RAT-free set S can always be extended by appending any new point P that does not lie

(i) on any circle having two points A, B of S as diameter, and

(ii) on the perpendiculars to AB through A and B (see Figure 127).

Thus there are RAT-free sets of every size having very general shapes.

The opposite of a RAT-free set would be one in which *every* subset of 3 points determines a right-angled triangle. Of course, for a set consisting of just 3 points this is trivially achieved by choosing the vertices of any right triangle. However, what about sets having more than 3 points? For $n = 4$, the vertices of a square constitute such a set—all 4 of its triangles are right-angled. For the set of 5 points consisting of the vertices and center of a square, 8 of its 10 triangles are right-angled, its only fail-

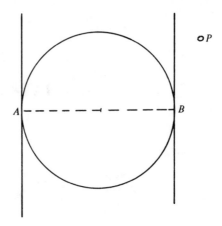

FIGURE 127

ures being the two collinear triads. As we shall see, it turns out that, for $n > 4$, there is no set S of n points which has *all* of its triangles right-angled; some set of 3 points must fail to determine a right-triangle. That is to say, every set S with $n > 4$ contains a subset T of 3 points which is RAT-free. Now the question we want to consider here is the size to which a RAT-free subset T can be extended.

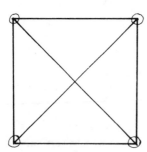
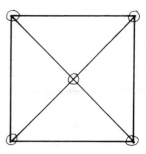

FIGURE 128

How big a RAT-free subset can one guarantee to be contained in every planar set S of n points?

The answer must apply to *all* sets of n points, not just a particular case; that is, we are asking for the size of the biggest RAT- free subset that one can *always count on finding* in S, no matter how the points of S may be scattered over the plane. This number, then, not depending on anything else, is just a function of the integer n; accordingly, let's denote it by $f(n)$. Because $f(n)$ is both a minimum and a maximum, it is very easy to lose sight of just exactly what it stands for. If the maximum size of a RAT-free subset were to be determined for each planar set S of n points, $f(n)$ would be the minimum of all these maxima. Again, $f(n)$ is the biggest minimum *that can be guaranteed* for the size of a RAT-free subset; as a guarantee, it is a minimum, but, curious as to how far we can go in building a RAT-free subset, we are not content to stop short of the maximum guarantee (i.e, the maximum minimum). These minimum-maximum quantities can be very confusing, but for our humble purposes we need not belabor the concept further.

Naturally we are most curious about how $f(n)$ depends on n, and our heart's desire is to derive a formula that describes this relationship. As so often happens with the many beautiful problems in combinatorial geometry, the object of our investigations seems to be very difficult to pin down exactly; in such cases, we are pleased to find any significant bounds that help us to close in on our elusive quarry.

Would you believe that in any set of n points there always exists a RAT-free subset which contains at least $\sqrt{n}$ points, i.e.,

$$f(n) \geq \sqrt{n}?$$

Accordingly, in *any* set of 50 points there is always some 8 points, none of whose $\binom{8}{3} = 56$ triangles is right-angled. Also, while all $\binom{n}{3}$ triangles of an n-set might fail to be right-angled, making the entire set RAT-free, we will show that one cannot always count on more than a RAT-free subset of size $2\sqrt{n}$, thus subjecting the universal number $f(n)$ to the bounds

$$\sqrt{n} \leq f(n) \leq 2\sqrt{n}.$$

And how delightful it is to find that the proof of these attractive results is nothing short of an absolute gem.

2. Proof of $\sqrt{n} \leq f(n)$

The problem of proving $\sqrt{n} \leq f(n)$ appears to be very difficult because we simply have no idea of the size of $f(n)$ for an arbitrarily chosen positive integer n. As boggling as it may seem to the intuition, it has often been found that an unknown function is sufficiently well behaved for us to infer useful facts about its general values from properties we are able to observe in the values it takes over a special subset of its domain. For the special subset $n = 5, 10, 17, \ldots, k^2 + 1, \ldots$, we shall give a beautiful proof that

$$f(k^2 + 1) \geq k + 1 \quad (k \geq 2);$$

from this result, and the obvious $f(n + 1) \geq f(n)$, we shall see that the desired $\sqrt{n} \leq f(n)$ follows easily *for all* n.

At the fundamental level, this is all another triumph for the famous pigeonhole principle. The following little result (due to Paul Erdős and George Szekeres in 1935), based on the pigeonhole principle, is, in turn, the basis of the rest of our proof in this section.

Theorem. *Let m and n be positive integers. Then, in any sequence $S = \{y_1, y_2, \ldots, y_{mn+1}\}$ of $mn + 1$ real numbers, there must exist either*

(i) *a nondecreasing subsequence $a_1 \leq a_2 \leq \cdots \leq a_r$ of length at least $m + 1$, or*
(ii) *a nonincreasing subsequence $b_1 \geq b_2 \geq \cdots \geq b_t$ of length at least $n + 1$.*

To prove this, let each term y_i in S be assigned a pair of "coordinates" (u_i, v_i) as follows: of all the nondecreasing subsequences of $S' = \{y_i, y_{i+1}, \ldots y_{mn+1}\}$ that *begin* with y_i, let u_i be the length of the longest; similarly, let v_i be the length of the longest nonincreasing subsequence of S' that *begins* with y_i.

For example, suppose $m = 3$, $n = 4$, and

$$S = \{6, 1, 4, 3, 21, 8, 4, 9, 6, 5, 11, 31, 4\};$$

then for $y_8 = 9$, we would have $u_8 = 3$ (from $\{9, 11, 31\}$), and $v_8 = 4$ (from $\{9, 6, 5, 4\}$), giving 9 the coordinates (3,4); similarly, $y_7 = 4$

would have coordinates (4,2). To each y_i, then, we have assigned a pair of positive integers (u_i, v_i).

Now, if any u_i is as great as $m + 1$, or any v_i is as great as $n + 1$, the desired conclusion follows. Suppose to the contrary that all $u_i \leq m$ and all $v_i \leq n$, that is, that

$$u_i \in \{1, 2, \ldots, m\} \quad and \quad v_i \in \{1, 2, \ldots, n\}.$$

In this case, the total number of *different* pairs of coordinates (u_i, v_i) that are possible is mn. But each of the $mn + 1$ terms y_i of S is assigned a pair (u_i, v_i). By the pigeonhole principle, some two of these pairs must be the same i.e., for some

$$i < j, \quad u_i = u_j \quad and \quad v_i = v_j.$$

However, we can easily show that this is impossible.

$$S : \cdots \quad \underset{(u_i, v_i)}{y_i} \quad \cdots \quad \underset{(u_j, v_j)}{y_j} \quad \cdots$$

Let $\{y_j \leq a_2 \leq a_3 \leq \cdots \leq a_{u_j}\}$ be a longest nondecreasing subsequence that begins with y_j. Now, if $y_i \leq y_j$, then $\{y_i \leq y_j \leq a_2 \leq \cdots \leq a_{u_j}\}$ is an even longer nondecreasing subsequence that begins with y_i, implying the contradiction $u_i > u_j$. In order to avoid this, it must be that $y_i > y_j$, which, by a similar argument, yields that $v_i > v_j$ (recall $i < j$). Thus we get a contradiction in either case and we are led to the conclusion that indeed either some

$$u_i \geq m + 1 \quad \text{or some} \quad v_i \geq n + 1.$$

In particular, for $m = n = k$, we have that in any sequence of length $mn + 1 = k^2 + 1$, there is a nondecreasing subsequence of length $k + 1$ or a nonincreasing subsequence of length $k + 1$.

Now let's go on to the proof of the result

$$k + 1 \leq f(k^2 + 1) \quad (k \geq 2).$$

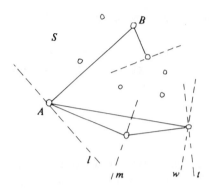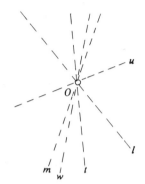

FIGURE 129

Let S be any set of $n = k^2 + 1$ points in the plane. For each pair of points (A, B) in S, let the **direction perpendicular** to AB be noted (these can be recorded by lines through a fixed point O; see Figure 129).

No matter how great k may be, the $\binom{k^2+1}{2}$ such directions still constitute only a *finite* set. Thus there is no difficulty in choosing a direction d that is *not* among these perpendiculars. Now then, let cartesian coordinates be assigned to the points of S relative to any frame of reference that has x-*axis in the chosen direction d*. Since the direction of the x-axis is not among the perpendiculars, no segment AB which joins two points of S will be parallel to the y-axis, that is, no two points will have the same x- coordinates.

The purpose of all this is to order the $k^2 + 1$ points of S according to strictly increasing abscissae:

$$(x_1, y_1), (x_2, y_2), \ldots, (x_{k^2+1}, y_{k^2+1}),$$

where

$$x_1 < x_2 < \cdots < x_{k^2+1}.$$

Such an ordering of the points automatically imposes some sequential order upon the corresponding ordinates, whatever their values might be:

$$y_1, y_2, y_3, \ldots, y_{k^2+1}.$$

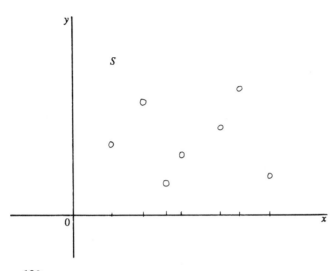

FIGURE 130

Appealing to our earlier result, with $m = n = k$, we see that there exists in this sequence of y's either

(i) a nondecreasing subsequence of length $\geq m + 1 = k + 1$, or
(ii) a nonincreasing subsequence of length $\geq n + 1 = k + 1$.

For definiteness, suppose the subsequence is

$$y_a, y_b, y_c, \ldots, y_r$$

and is nondecreasing. (The case of a nonincreasing subsequence is exactly similar.) Thus we have in hand a subset T of at least $k + 1$ points

$$T = \{(x_a, y_a), (x_b, y_b), \ldots, (x_r, y_r)\},$$

where

$$x_a < x_b < \cdots < x_r \qquad \text{and} \qquad y_a \leq y_b \leq \cdots \leq y_r.$$

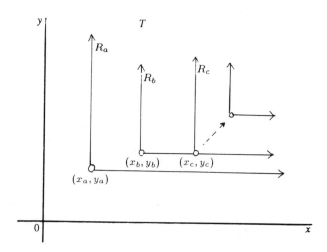

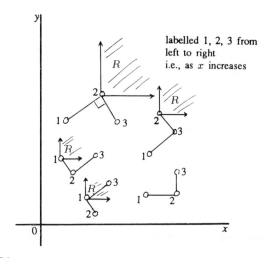

FIGURE 131

That is to say, at any of these points P of T, all the points farther along the subsequence T lie in a region R which is to the *right* of P and *at the same level or above P* (see Figure 131).

It is not difficult to see that no 3 of these points could determine a right angle; for, consider the arms of such a prospective right angle:

(i) if one arm is parallel to the x-axis, then the other is parallel to the y-axis in defiance of our choice of direction d for the x-axis; and

(ii) otherwise, one arm has positive slope, requiring the other to have negative slope (the product of the slopes is -1), which always places one of the points outside of its acceptable region R.

We conclude then that the subset T, consisting of at least $k + 1$ points of S, is indeed a RAT-free subset. Since this is valid for all sets S with $n = k^2 + 1$ points, we have that

$$f(k^2 + 1) \geq k + 1, \quad \text{as desired.}$$

Now to the easy completion of the proof of this section.

Let n be an arbitrary positive integer. Whatever its value, it is either a perfect square or it lies between two consecutive perfect squares. In any case, there is a unique positive integer k such that

$$(k - 1)^2 < n \leq k^2,$$

i.e.,

$$(k - 1)^2 + 1 \leq n \leq k^2.$$

Now, if a set of n points can always be counted upon to contain a RAT-free subset of size $f(n)$, then any set of $n + 1$ points, containing as it does a set of n points within it, can surely do as much, and we have

$$f(n + 1) \geq f(n).$$

Since $n \leq k^2$, then $\sqrt{n} \leq k$. Thus we have altogether that

$$f(n) \geq f(n - 1) \geq \cdots \geq f\left((k - 1)^2 + 1\right) \geq k \geq \sqrt{n},$$

i.e.,

$$f(n) \geq \sqrt{n},$$

as claimed.

3. Proof of $f(n) \leq 2\sqrt{n}$

This time the desired result is almost equivalent to the property

$$f(k^2) \leq 2k - 2 \quad (k \geq 2),$$

which is very nicely proved as follows.

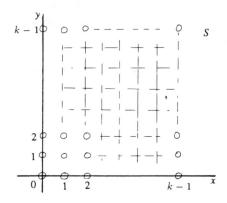

Consider a $k \times k$ square of lattice points; for example, the set (x, y) where $0 \leq x, y \leq k - 1$. Clearly this constitutes a set of $n = k^2$ points in the plane. Let X be any RAT-free subset of S, and let P be any point of X. Now consider any other points of X that lie in P's row and P's column. If P is the *only* point of X in its row, then color it red; if P is the *only* point of X in its column, color it blue. One might wonder at this point whether every point of X gets colored by this procedure. Consider then the possibility that some point Q of X is neither red nor blue. In this case it is not alone in either its row or its column, and it is therefore the vertex of an impossible right triangle determined by 3 points of X (Figure 133). If anything, a problem might arise in the case of X being colored *both* red and blue. This is certainly possible (if it is alone in both its row and column), and to forestall any problem due to this, let us agree to color such points either red or blue, but not both. Finally, let r be the total number of red points and b the total number of blue ones; then the size of X is $r + b$.

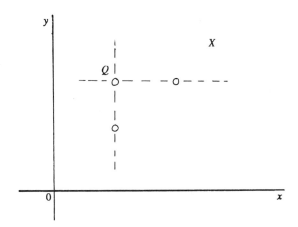

FIGURE **133**

Now, if r were to be as great as k, each of the k rows of S would have to contain a single red point and nothing else, implying that these k red points would constitute all of X, and

$$\text{the cardinality of} \quad X = |X| = r + b = k + 0 = k.$$

Similarly, if $b = k$, there could be no red points in X, and

$$|X| = r + b = 0 + k = k.$$

Because we are interested only in $n \geq 4$, we have $n = k^2 \geq 4$, and $k \geq 2$. In this case, it follows easily that

$$k + k \geq 2 + k, \quad \text{and} \quad |X| = k \leq 2k - 2.$$

Since there are only k rows and k columns in S, if neither $r = k$ nor $b = k$, then both

$$r \leq k - 1 \quad \text{and} \quad b \leq k - 1,$$

giving

$$|X| = r + b \leq 2(k-1) = 2k - 2.$$

Thus, in all cases we have

$$|X| \leq 2k - 2,$$

and we conclude that no RAT-free subset X of such a set of k^2 points P can contain more than $2k - 2$ points. Accordingly, our guarantee of a RAT-free subset of size $f(k^2)$ must be bounded by this ceiling, and we have

$$f(k^2) \leq 2k - 2.$$

Finally, let n be an arbitrary positive integer. Whatever its value, there is a unique positive integer k such that

$$k^2 \leq n < (k+1)^2,$$

i.e.,

$$k^2 \leq n \leq (k+1)^2 - 1.$$

Accordingly, $k \leq \sqrt{n}$, and

$$f(n) \leq f\left((k+1)^2 - 1\right) \leq f\left((k+1)^2\right) \leq 2k \leq 2\sqrt{n},$$

giving the desired

$$f(n) \leq 2\sqrt{n}.$$

In writing up this morsel, extensive use was made of the excellent article "On the Largest RAT-FREE Subset of a Finite Set of Points" by Wah Keung Chan, a student at McGill University, Montreal, which appeared as the lead article in the outstanding journal $\pi\mu\epsilon$, Spring 1987. Credit is given there to H. L. Abbott [1] for the application of the Erdős-Szekeres result to the proof of $f(n) \geq \sqrt{n}$, and to A. Seidenberg [2] for the proof of $f(n) \leq 2\sqrt{n}$.

REFERENCES

1. Abbott, H. L., On a conjecture of Erdős and Silverman in combinatorial geometry, *Journal of Combinatorial Theory*, Series A 29 (1980), No. 3, 380–381.

2. Seidenberg, A., A simple proof of a theorem of Erdős and Szekeres, *Journal London Math. Soc.* 34 (1959) 352.

MORSEL **57**

A FURTHER NOTE ON OLD MORSEL 23

In *Mathematical Morsels* (Dolciani Series, Vol. 3) the 23rd problem is the following (page 48–51):

> Given $2n + 3$ points in the plane, no 3 on a line and no 4 on a circle, prove that it is always possible to find a circle C that goes through 3 of the given points and splits the other $2n$ in half, that is, has n on the inside and n on the outside.

Two solutions are given in *Mathematical Morsels* and a third one in *Mathematical Gems III* (Dolciani Series, Vol. 9, pp. 18–19). In each case, passing reference is made to the *number* of such solution- circles C that must occur and a rough lower bound is given. These bounds are extremely loose, however, and it is now my pleasure to present a vastly improved bound that is due to Karel Post of the University of Technology, Eindhoven, The Netherlands. As with so many things, this result follows from little more than a simple observation, once one has maneuvered himself into a position to see it. Our goal in this section is to show that

the number of solution-circles is always at least $\dfrac{(2n + 3)(n + 1)}{3}$.

To set the stage for our brief encounter, and as a little bonus, we begin with Professor Post's neat proof that a set of 5 points (i.e., $n = 1$) always provides precisely 4 solution-circles.

Solution

We follow the lead taken by my friend and colleague Lee Dickey in his inversive solution, given in *Mathematical Morsels*. An understanding of his work is not assumed in what follows.

Let $S = \{P, Q, R, X, Y\}$ be a set of $2n + 3 = 5$ points in the plane, no 3 on a line and no 4 on a circle. Each 3-subset of these points, say $\{P, Q, R\}$, not being on a line, determines a nondegenerate circle, and since no 4 points of S lie on a circle, P, Q, and R are the only points of S that lie on this circle. That is to say, each of the $\binom{5}{3} = 10$ triples of the given points determines a unique circle. The number of these circles through a specified point P of S, then, is the number of triples from S which contain P, that is $\binom{2n+2}{2} = \binom{4}{2} = 6$, the number of ways you can pick a pair $\{Q, R\}$ from S to go with P. So, there are 6 circles through each of our 5 given points of S. It would be nice to know how many of the 6 circles through P are point-splitting and how many are not. As a first step in investigating the possibilities in this area, we invert the set S in any circle I having center P. This carries P to the point at infinity and yields a set S' of finite images containing $2n + 2 = 4$ points.

Now let us be ready to move freely back and forth between the worlds of the antecedents $S = \{P,\ Q,\ R,\ X,\ Y\}$ and the images $S' = \{Q',\ R',\ X',\ Y'\}$. Clearly, PQR is one of the 6 circles through P. In S', the image of circle PQR is the straight line $Q'R'$ (containing the image of P at infinity). Now the central point of our approach is that the circle PQR splits the remaining two given points X and Y if and only if its image $Q'R'$ is a *line* that splits the other two images X' and Y'; this follows from the fact that the images of all points inside PQR lie on one side of $Q'R'$ (like X' in the figure), and the images of all the points outside PQR lie on the other side of $Q'R'$ (e.g., Y'). Thus the inversion has transformed our problem into the determination of how many of the $\binom{2n+2}{2}$ straight lines determined by S' split the remaining points of S' in half, 2 on the line itself and n on each side; henceforth, for brevity, let us refer to such a line simply as an image-splitting line, it being understood that only *equal* splits are intended.

Next we need to digress briefly to show that S' enjoys the same qualifying properties that S does, namely

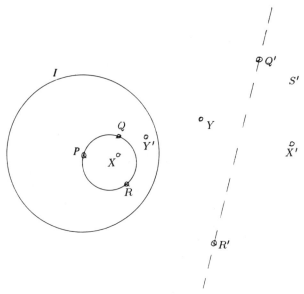

FIGURE 134

(a) no 3 on a line,
(b) no 4 on a circle.

This is easily accomplished by showing that violations quickly yield contradictions. Suppose Q', R', and X' lie on a line L. Then either L goes
through the center of inversion P or it doesn't. If it does, then we have
the points Q, R, and X of S also on L (not to mention P); if it doesn't,
then the points P, Q, R, and X all lie on a circle in S. Four concyclic
points in S' yield similar contradictions in S.

Now then, S' contains only 4 points, and because no 3 lie on a line,
there are only two possible configurations that S' can assume:

 (i) S' determines a convex quadrilateral,
(ii) S' consists of one point inside the triangle formed by the other
      three.

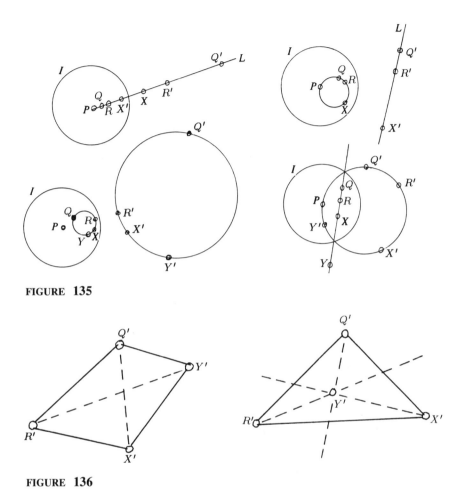

FIGURE 135

FIGURE 136

Whatever the case, a side of the convex hull (that is, the quadrilateral in (i), and the triangle in (ii)), is *not* a point-splitting line in S', implying that its antecedent circle is not point-splitting in S, either. In fact, of the 6 lines in (i), only the two "diagonals" are image-splitting, and we conclude that only 2 of the 6 circles through P are point-splitting solution-circles. Similarly, in (ii), 3 lines split the images and 3 don't, giving 3 solution-circles through P.

Summarizing, then, through any point P of S, at least 2 of the 6 circles are solution-circles and at least 3 are not (the 6th circle may be of either kind).

Now the circle PQR is counted among the 6 through P, and also among the sets through Q and R. Therefore, in counting up the solution-circles for all 5 points of S, each circle is counted exactly 3 times. Since each of the 5 points contributes at least 2 to the total, it must amount to at least 10 altogether. With each circle counted 3 times, a total of 10 or more would require at least $10/3$ *different* circles. Thus we have that the number of solution-circles must be at least 4.

Dealing similarly with the circles that are *not* solution-circles, we obtain, at 3 or more per point of S, a total of at least 15, requiring at least $15/3 = 5$ different nonsolution-circles. Therefore, of the 10 circles determined by S, at least 4 *are* solution-circles and at least 5 *are not*. Since there remains only one circle up for grabs, the final count must be either 4 solution-circles and 6 nonsolution- circles or 5 of each kind. We shall see that the former must hold by showing that the latter leads to a contradiction.

Accordingly, suppose there are 5 circles of each kind. Clearly the 5 points of S go together into $\binom{5}{2} = 10$ pairs (P, Q). We shall show that the number of solution-circles that go through *both* members of a pair (P, Q) is always either 1 or 3. As above, let S be inverted in any circle having center P. In S', then, a circle through P and Q becomes a line through Q', and a solution-circle PQR is distinguished by the fact that its image $Q'R'$ is an image-splitting line in S'. Considering all 8 of the possible images in the case (i) and (ii), it is clear from the figures that every image point lies on either 1 or 3 image-splitting lines in S', establishing the desired conclusion.

Thus, let us count up the solution-circles for all 10 of the point-pairs (P, Q). Suppose that p of the pairs contribute 1 to our total and that q of them contribute 3. Each of the 10 pairs does one or the other, and we have

$$p + q = 10.$$

Now a solution-circle (A, B, C) would be counted exactly 3 times in this sum—once for each of the pairs $(A, B), (B, C), (C, A)$. Altogether, then, the 5 of them must yield a grand total of 15 for the count, and we have

$$p + 3q = 15.$$

Solving these equations gives $q = 2\frac{1}{2}$, contradicting its integral character.

Hence a set of 5 solution-circles is an untenable hypothesis and we conclude that a set of 5 points always provides exactly 4 solution-circles.

Our main result is now just a corollary to this powerful inversive approach. If S contains $2n + 3$ points, then S' contains $2n + 2$ points. The crucial thing we need to prove is that, for each image point Q', at least one of its lines $Q'R'$ is an image-splitting line in S' (recall that image-splitting refers only to *equal* splits). Momentarily assuming that this is true, the desired conclusion is reached quickly as follows. An image-splitting line $Q'R'$ in S' means, of course, a point-splitting solution-circle PQR in S. If there exists at least one solution-circle PQR for each of the $2n + 2$ image points Q', a collection of at least $2n + 2$ circles would result. However, these circles go together in identical pairs. If Q' pairs with R' to determine an image-splitting line, then the same line arises when considering R', and a duplicate circle results.

Thus we can claim at least $(2n + 2)/2 = n + 1$ different solution-circles in our set. But each of these circles goes through P. Therefore, we conclude that there are at least $n+1$ different solution-circles through each point of S. Counting up over S, we get a total of at least $(2n + 3)(n + 1)$. Since each circle is counted exactly 3 times, we have that S must contain at least $(2n+3)(n+1)/3$ different solution-circles.

Our story is completed with the easy proof that each image point Q' must lie on at least one image-splitting line $Q'R'$.

Let a directed line L sweep around the plane by being spun about Q'. Since no 3 images of S' lie on a line, L will encounter the points of S' *one at a time*. In each position that L passes through a second point R' of S', let the number of points of S' on each side of L be recorded. Suppose in the initial position there are m images on the right side of L and $m - k$ on the left, where $k \geq 0$. Then $m + (m - k) = 2n$, yielding k to be some even integer $2r$. In these terms, the right side has m images and the left side $m - 2r$.

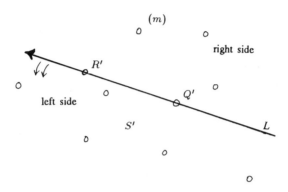

FIGURE 137

Because L meets the points of S' one at a time, the number of points on a side of L, from one position to the next, must change by a small increment; in fact, it can only change by 1, 0, or −1 (a little experimenting will quickly convince you of this). Now, after spinning halfway around Q', the right and left sides of L have changed places, and accordingly the number of points on the right side has dropped from m to $m - 2r$. While the number of images on the right side might fluctuate as L spins around, there is no avoiding its passing through all the intermediate integers between any two of its values. Hence at some position, L must have had $m - r$ images on its right side, i.e., exactly half the total number of image points $m + (m - k) = m + (m - 2r) = 2m - 2r$, other than the two on L itself.

REFERENCES

All references, except for a very few that are specifically noted, are to the journal *Crux Mathematicorum,* published by the Canadian Mathematical Society.

Morsel 1. 1979, 102, Olympiad Corner 4, Sample Problem from the Mathematics Student.

Morsel 2. 1983, 56, Problem 712, proposed by Donald Aitken, Northern Alberta Institute of Technology, Edmonton, Alberta; solved by Bernard Baudiffier, College de Sherbrooke, Sherbrooke, Quebec.

Morsel 3. 1980, 17, Problem 418, proposed by James Gary Propp, student, Harvard College; solved by Leroy F. Meyers, The Ohio State University.

Morsel 4. 1986, 58, Problem 995, proposed by Hidetosi Fukagawa, Yokosuka High School, Tokai-City, Aichi, Japan; solved by Sam Baethge, San Antonio, Texas.

Morsel 5. 1981, 185, Problem 555, proposed by Michael W. Ecker, Pennsylvania State University, Worthington Scranton Campus; solved by John T. Barsby, St. John's-Ravenscourt School, Winnipeg, Manitoba.

Morsel 6. 1983, 221, Problem 759, proposed by Jack Garfunkel, Flushing, New York; solved by Leon Bankoff, Los Angeles, California; comment by Leroy F. Meyers, The Ohio State University.

Morsel 7. 1984, 157, Problem 817, solved by Stanley Rabinowitz, Digital Equipment Corporation, Merrimack, New Hampshire; solved by Murray Klamkin and A. Meir, University of Alberta.

Morsel 8. 1980, 188, Problem 466, proposed by Roger Fischler, Carleton University, Ottawa; solved by Gregg Patruno, student, Stuyvesant High School, New York.

Morsel 9. 1976, 80–81, Problem 108, proposed by Viktors Linis, University of Ottawa; solved by F. G. B. Maskell, Algonquin College.

Morsel 10. 1982, 325, Problem 705, proposed by Andy Liu, University of Alberta; solved by Leroy F. Meyers, The Ohio State University.

Morsel 11. 1976, 95, Problem 111, proposed by H. G. Dworschak, Algonquin College; solved by Leo Sauvé , Algonquin College.

Morsel 12. 1977, 143, Problem 206, proposed by Dan Pedoe, University of Minnesota; solved by Dan Sokolowsky, Yellow Springs, Ohio.

Morsel 13. Part (b): 1977, 73, Problem 188, proposed by Daniel Rokhsar, Susan Wagner High School, Staten Island, New York; solved by W. J. Blundon, Memorial University of Newfoundland.

Morsel 14. 1982, 212, Problem 657, proposed by Hgo Tan, student, J. F. Kennedy High School, The Bronx, New York; solved by J. T. Groenman, Arnhem, The Netherlands.

Morsel 15. 1979, 233, Problem 396, proposed and solved by D. Bernshtein in the Russian journal *Kvant* (found there by Viktors Linis, University of Ottawa, who submitted it to *Crux Mathematicorum*).

Morsel 16. 1982, 294, Problem 686, proposed by Charles W. Trigg, San Diego, California; solved by Jordi Dou, Barcelona, Spain.

Morsel 17. 1979, 271, Problem 405. This was found by Viktors Linis, University of Ottawa, in the Russian journal *Kvant* (No. 8, 1977, 46, Problem M419), who submitted it to *Crux Mathematicorum*. The solution, given in *Kvant,* is by I. Klimova.

A write-up of this problem, similar to the one given here, was included in my guest column of "Mathematical Games" (*Scientific American,* August, 1980, 15) when I substituted for Martin Gardner.

Morsel 18. 1982, 113, Problem 626, proposed and solved by A. Liu, University of Alberta.

Morsel 19. 1984, 122, Problem 807, proposed by D. J. Smeenk, Zaltbommel, The Netherlands; solved by Richard Rhoad, New Trier High School, Winnetka, Illinois.

Morsel 20. 1978, 21, Problem 246, proposed by Kenneth M. Wilke, Topeka, Kansas; solved by Richard A. Gibbs, Fort Lewis College, Durango, Colorado.

Morsel 21. 1981, 127, Problem 540, proposed by Leon Bankoff, Los Angeles, California; solved by J. T. Groenman, Arnhem, The Netherlands.

Morsel 22. 1976, 183, Problem 147, proposed by Steven R. Conrad, Benjamin N. Cardozo High School, Bayside, New York; solved by Leon Bankoff, Los Angeles, California.

Morsel 23. 1979, 171, Problem 380, proposed and solved by G. P. Henderson, Campbellcroft, Ontario.

Morsel 24. 1979, 84, Problem 358, proposed and solved by Murray Klamkin, University of Alberta.

Morsel 25. 1977, 165, Problem 212, proposed by Bruce McColl, St. Lawrence Collegiate, Kingston, Ontario, Canada; solved by Doug Dillon, Brockville, Ontario, and, independently, by Daniel Flegler, Waldwick High School, Waldwick, New Jersey.

Morsel 26. 1979, 104, Olympiad Corner 4, Sample Problem from the *Two-Year College Mathematics Journal,* 1977 (now called *The College Mathematics Journal*).

Morsel 27. 1976, 226, Problem 161, proposed by Viktors Linis, University of Ottawa; solved by Hippolyte Charles, Waterloo, Quebec.

Morsel 28. 1979, 273, Problem 406, proposed by W. A. McWorter Jr.; solved by E. J. Barbeau, University of Toronto.

Morsel 29. Parts (i) and (ii): 1975, 19–20, Problems 8 and 9: posed by Jacques Marion, University of Ottawa; part (i) solved by André Ladouceur, Ecole Secondaire De La Salle; part (ii) solved by the Proposer.

Part (iii): 1985, 47–48, Problem F.2436, Olympiad Corner 62, taken from the Russian journal *Kozepiskolai Matematikai Lapok* 67 (1983), 80; solved by Gali Salvatore, Perkins, Quebec.

Morsel 30. 1976, 202, Problem 159, proposed by Viktors Linis, University of Ottawa; solved by Clayton Dodge, University of Maine at Orono.

Morsel 31. 1978, 161, Problem 294, proposed by Harry D. Ruderman, Hunter College, New York; solved by F. G. B. Maskell, Algonquin College, Ottawa.

Morsel 32. 1977, 252, Problem 189, proposed by Kenneth S. Williams, Carleton University; solved by Basil Rennie, James Cook University.

Morsel 33. 1985, 33, Problem 888, proposed by W. J. Blundon, Memorial University of Newfoundland; solved by Kenneth M. Wilke, Topeka, Kansas.

Morsel 34. 1985, 327, Problem 973, proposed by Loren C. Larson, St. Olaf College, Northfield, Minnesota; solved by Walther Janous, Ursulinengymnasium, Innsbruck, Austria.

Morsel 35. 1984, 131, Problem 814, proposed and solved by Leon Bankoff, Los Angeles, California.

Morsel 36. 1981, 31, Problem 510, proposed by Gali Salvatore, Perkins, Quebec; solved by W. J. Blundon, Memorial University of Newfoundland.

Morsel 37. 1980, 58, Problem 433, proposed by Dan Sokolowsky, Antioch College, Yellow Springs, Ohio; solved by Clayton W. Dodge, University of Maine at Orono.

Morsel 38. 1976, 203, Problem 160, proposed by Viktors Linis, University of Ottawa; solved by André Ladouceur, Ecole Secondaire De La Salle.

Morsel 39. 1986, 56, Problem 993, proposed by Walther Janous, Ursulinengymnasium, Innsbruck, Austria; solved by Murray Klamkin, University of Alberta.

Morsel 40. 1982, 85, Problem 619, proposed by Robert A. Stump, Hopewell, Virginia; solved by Kesiraju Satyanarayana, Gagan Mahal Colony, Hyderabad, India.

Morsel 41. 1982, 82, Problem 617, proposed by Charles W. Trigg, San Diego, California; solved by W. J. Blundon, Memorial University of Newfoundland.

Morsel 42. 1977, 70, Problem 185, proposed by H. G. Dworschak, Algonquin College; the solution in part (b) by T. J. Griffiths, A. B. Lucas Secondary School, London, Ontario, Canada.

Morsel 43. 1978, 198, Problem 307, proposed by Steven R. Conrad, Benjamin Cardozo High School, Bayside, New York; solved by Kenneth S. Williams, Carleton University, Ottawa.

Morsel 44. 1982, 319, Problem 698, proposed by Hippolyte Charles, Waterloo, Quebec; solved by Leroy F. Meyers, The Ohio State University.

Morsel 45. 1979, 83, Problem 357, proposed by Leroy Meyers, The Ohio State University; solved by Gilbert W. Kessler, Canarsie High School, Brooklyn, New York.

Morsel 46. 1979, 227, Olympiad Corner 8, Problem 5 of the British Mathematical Olympiad, 1979.

Morsel 47. 1983, 128–131, Some Boolean inequalities from a triangle, by J. L. Brenner.

Morsel 48. 1980, 188, Problem 465, proposed by Peter A. Lindstrom, Genesee Community College, Batavia, New York; solved by Andy Liu, University of Alberta.

Morsel 50. 1982, 256, Problem 674, proposed by George Tsintsifas, Thessaloniki, Greece; solved by Howard Eves, University of Maine.

Morsel 52. 1975, 32–33, Problem 19, proposed by H. G. Dworschak, Algonquin College; solved by Leo Sauvé, Algonquin College.

Morsel 53. 1982, 211, Problem 656, proposed by J. T. Groenman, Arnhem, The Netherlands; solved by Jordi Dou, Barcelona, Spain.